Food Policy Reforms in Central Asia: Setting the Research Priorities

EDITED BY
SURESH BABU AND ALISHER TASHMATOV

International Food Policy Research Institute
Washington, D.C.

Library of Congress Cataloging-in-Publication Data
Food policy reforms in Central Asia : setting the research priorities / edited by Suresh Babu and Alisher Tashmatov:

p. cm.

Papers presented at a regional workshop.
Includes bibliographical references.
ISBN 0-89629-702-0 (pbk)
1. Agriculture and state—Asia, Central—Congresses. 2. Food Supply—Government policy—Asia, Central—Congresses. 3. Agriculture—Research—Asia, Central—Congresses.
I. Babu, Suresh Chandra. II. Tashmatov, Alisher.

HD2065.22.Z8 F66 2000
333.1'858—dc21

00-054271

Contents

Tables

Figures

Foreword

Since the break-up of the Soviet Union, the countries of the Central Asia region have undertaken varying levels of policy reforms in their food, agricultural, and natural resources sectors. Despite these reforms, available evidence indicates that poverty and natural resource degradation are on the rise in this region. While this situation poses enormous challenges to policymakers, it also provides opportunities for redefining the process of economic and political reform.

How well the policymakers use these opportunities will depend on the availability of information on the impact of various policy alternatives on the living standards of the people. Such information continues to be meager in this region, and policy research in the food, agricultural, and natural resource sectors is at a rudimentary stage in these countries. This volume brings together a group of policy researchers from Central Asia and elsewhere to address the emerging food policy issues, to identify the information gaps, and to set priorities for policy research.

"Food Policy Reforms in Central Asia: Setting the Research Priorities" is an outcome of the initial collaboration between the International Food Policy Research Institute (IFPRI) and the institutions involved in analyzing and advising policymakers on food policy reforms in the Central Asia region. The chapters of this volume represent the presentations and group discussions from the regional workshop held in Tashkent in July 1999 entitled "Food, Agriculture, and Natural Resource Policy Research in Central Asia: Setting the Priorities." The workshop was jointly organized by the Uzbekistan Ministry of Agriculture and Water Resources, the Rural Restructuring Agency, the Tashkent State Agrarian University, and IFPRI. The Central Asia and the Caucasus (CAC) program of the Consultative Group on International Agricultural Research (CGIAR) provided both financial and logistical support for the implementation of the workshop.

The workshop and this volume mark the beginning of IFPRI's research and outreach commitment to Central Asia. The papers of this volume identify key issues that need immediate research attention from those involved in the food policy reform process in the region. The workshop identified market reforms, regional trading arrangements, agribusiness and postharvest technology, food security and agricultural diversification, water use management, and sustainable use of rangelands as the priority policy research themes.

It is our hope that the papers of this volume and the analysis contained therein will provide new insights and inform those involved in the policy reform process of the Central Asia region.

Per Pinstrup-Andersen
Director General
International Food Policy
Research Institute

Abduvokhid Juraev
Deputy Minister
of Agriculture and Water Resources
Republic of Uzbekistan

Acknowledgments

This volume contains papers that were presented at a three-day workshop entitled "Food, Agricultural, and Natural Resource Policy Research in Central Asia: Setting the Priorities," held in Tashkent, Uzbekistan in July 1999. The papers that were prepared in Russian by collaborators in Central Asia have been translated into English for this volume.

The workshop was co-organized by the International Food Policy Research Institute (IFPRI), the Uzbekistan Ministry of Agriculture and Water Resources, the Rural Restructuring Agency, and the Tashkent State Agrarian University. The workshop was cosponsored by the Central Asia and the Caucasus (CAC) program of the Consultative Group on International Agricultural Research (CGIAR) and IFPRI.

Many people contributed to the success of the workshop and to the production of this volume. We are thankful to the Honorable Abduvokhid Juraev, Deputy Minister for Agriculture, Uzbekistan, and Per Pinstrup-Andersen, Director General of IFPRI, for their commitment to the agricultural development of Central Asia and for opening the workshop. Dr. Farruk Aknazarov, Director General of the Rural Restructuring Agency, was helpful in the organization of the workshop. Dr. H. Buriev, Rector of Tashkent State Agrarian University was instrumental in hosting the workshop.

We are grateful to the CAC regional office in Tashkent for their ground support in organizing the workshop, including travel arrangements and translation facilities. Without the committed help of Drs. Suren Beniwal, Mekhlis Suleimenov, and Zahir Khalikulov, the workshop would have faced many hurdles. Thanks are also due to Chandra Ranade, Jit Srivastava, and Mohinder Mudahar for their encouragement and support in the workshop.

Thanks are also due to Francesco Goletti, Rajul Pandya-Lorch, Peter Oram, and Erika Meng for their participation and help during the workshop; to Will Reidhead and Zakir Khalikulov who helped with translation and the organization of this volume; and to Heidi Fritschel, Evelyn Banda, Klaus von Grebmer, Lawrence Haddad, and Don Lippincott for their editorial support and guidance in the preparation of this volume.

We are also grateful to the 30 workshop participants from Kazakhstan, Kyrgyzstan, Tajikistan, Turkmenistan, and Uzbekistan for their contributions and preparation of papers under a tight deadline. It is the commitment of these individuals that made the workshop and this volume possible.

Keynote Address

PER PINSTRUP-ANDERSEN, Director General, IFPRI

Mr. Chairman, Honorable Deputy Minister for Agriculture, Dr. Juraev, the Director General of Farm Restructuring Agency, Dr. Aknazarov, the Director General of the Rural Restructuring Agency, Dr. Khusanov, the participants of the workshop, ladies, and gentlemen. Let me first of all welcome the esteemed participants of the workshop from all the five countries of Central Asia.

I am very glad to be here at the Tashkent State Agrarian University, an internationally renowned institution that has trained many agriculturalists from a number of developing countries over the years. I am particularly excited about the opportunity for the International Food Policy Research Institute (IFPRI) to collaborate with this university and the Uzbekistan Ministry of Agriculture and Water Resources in organizing this regional workshop on "Food, Agriculture, and Natural Resource Policy Research in Central Asia: Setting the Priorities."

Central Asia is a region that is facing enormous challenges in terms of food insecurity and natural resource degradation. But these challenges also provide opportunities for action that will result in improvements in the welfare of the population. Converting these opportunities into concrete policy actions is fundamental for making sustained progress.

Even though the world produces enough food for every person on earth, more than 800 million people are still hungry. About 500 million of them live in the Asia-Pacific region. As you know, except for Kazakhstan, all other Central Asian countries are designated as low-income food-deficit countries. These countries are particularly vulnerable to food insecurity and to fluctuations in global food prices. Thus, sustainable increases in agricultural productivity in Central Asia are essential for short-term poverty reduction and long-term food security.

Increasing agricultural productivity by small-holder farmers in Central Asia requires increased investments in agricultural research that will generate new and improved technologies. In addition to agricultural technology, improved institutions are also critical to help farmers obtain modern inputs and sell their outputs. An appropriate policy environment and its interaction with technology and institutions are also necessary.

This workshop aims to bring out the key issues that are relevant for creating an appropriate policy environment in the food, agriculture, and natural resources sectors. It will also identify the information gaps in addressing these policy issues and will derive a set of priorities for policy research.

At IFPRI, we are extremely fortunate to collaborate with the key policy researchers and analysts from the Central Asian countries. Once again, on behalf of IFPRI and myself, let me welcome the distinguished participants and thank the Uzbekistan Ministry of Agriculture and Water Resources and the Tashkent State Agrarian University for hosting this workshop. I wish you all the best in your deliberations in the three days to come.

Opening Address

ABDUVOKHID JURAEV
Deputy Minister of Agriculture and Water Resources
Republic of Uzbekistan

Dear distinguished guests, friends, ladies, and gentlemen:

I am pleased to welcome you to the rich and hospitable land of Uzbekistan. In my opinion there is no need to tell you how important and urgent this workshop is for the Central Asian countries.

The issues of food security and pursuing appropriate policies in this area are of great importance for the former Soviet Union countries under the present conditions of demographic growth and a complicated period of transition to a market economy. What are the priorities in agricultural development, in natural resource utilization, and in resolving population and food security issues? These issues are actually interrelated and complex and need to be studied and understood in a holistic way to identify appropriate solutions.

Sustainable development of agriculture is necessary to achieve sustained progress in social development and is also important for the sustainable use of natural resources and for providing food security for the growing population. Agriculture plays a leading role in the economy of the Republic of Uzbekistan. More than 60 percent of the population lives in rural areas; about 40 percent of the working population is employed in agricultural production; and 25 to 30 percent of the gross domestic product and 55 percent of hard currency earnings are received from agriculture. Most industrial production plants in the republic are in some way related to the agricultural sector.

As was stated by Mr. Islam Karimov, President of the Republic, the principle "The state is the main reformer and conductor of the reforms" is the most important of the five principles of the construction of the legal state. Considering the importance of agriculture in the economy since independence, the government of the Republic has adopted an intensive path of reforms in this sector. A program of "Deepening Economic Reforms in Agriculture" for the period 1998–2000 was adopted in 1998 by the president's decree. A special resolution on the implementation of this decree was adopted as well. Furthermore, the "Land Code" and the laws "On Agricultural Cooperatives" (*shirkat* farms), "On Farming Units," and "On Individual Farming Units" have been enacted by the Parliament of the Republic.

The intent of these documents is to deepen the economic reforms by increasing the number of farm owners and forms of ownership; by expanding their rights and control of the outputs; and by the rational utilization of land, water, and other resources.

According to the State Program of Development, by the year 2000 all collective farms will be transformed into cooperative ones, and the subsidiary farms will be transformed into individual farming units.

The Program of Deepening Reforms in Agriculture has been adopted in accordance with the No. 20 Resolution of the Cabinet of Ministers, dated March 19, 1998, which is aimed at the provision of sustainable development of this sector. As it was pointed out by President Karimov,"Assessing the adopted programs, we are imbued with the idea that these programs are the pledge, I would say guarantee, of our sustainable and dynamic progress along the way of reform and renewal of our society."

Significant reduction of the deficit in the supply of major products—meat, milk, vegetables, and fruit—has been achieved as the result of the reforms undertaken in the republic since independence. These reforms were aimed at improving living conditions for rural people. For example, distribution of land plots for the development of households is a key contribution of the reform process. Excess production is being delivered to both city and local farmers' markets.

Great attention in the Republic has been paid to attaining grain independence and self-sufficiency in vitally important kinds of food products and raw materials for the food industry. Thus, since 1996, the import of grains has been reduced significantly, and grain self-sufficiency has been achieved.

Reforming the system of agricultural finance has been carried out since independence. A new credit and finance institution, *Tadbirkiorbank*, was established to provide financing for developing new households and individual farming units, for small and medium-size enterprises, and for loans from the Small and Medium Enterprises Support Fund. The privatization of *Agroprombank* has been undertaken, followed by its division into three banks, each of which specializes in certain crops.

In Uzbekistan, as well as in the other postsocialist countries, the agricultural sector faces several problems. Specifically, the low efficiency in agricultural production, the lack of efficiency in the operation of the credit and finance system, the need for improvements to the processing industry of the agroindustrial sector, ecological issues, and the production of export-oriented commodities conforming to world standards continue to pose challenges.

There are no universal recipes for attaining the sustainable and humane development of agriculture and resolving the problems of food security given the varying economic, geographic, demographic, historic, and other characteristics of each country. In-depth studies and research are needed to resolve these issues in Uzbekistan. We appreciate the efforts of the International Food Policy Research Institute in arranging this workshop, which is highly relevant at the present stage of development in our country. Furthermore, we hope for the further support of this Institute as well as for the assistance of other similar international organizations with experience in the areas of agriculture, food security, and sustainable ecological development for the improvement and the well-being of our Republic and all the Central Asian countries.

The government of the Republic of Uzbekistan believes that the results of this workshop will contribute to forming a strategy for food security policy in the countries of this region and to establishing mutually beneficial contacts for resolving these issues by specialized regional and worldwide institutions and departments.

I wish all the participants of the workshop a successful and fruitful deliberation toward achieving the goals that have been set forward. I expect that our cooperation will last for a long time and the results of our collaboration will serve to further improve the well-being of the peoples of our countries.

Thank you for your attention.

PART I

Introduction

1 Introduction and Overview

SURESH BABU AND ALISHER TASHMATOV

Since independence, the countries of the Central Asia region have undergone a series of transitions from centrally planned economies to market-orientated systems. Despite great efforts by the five Central Asian countries and the external advice and efforts of international and bilateral agencies to help them follow a dynamic growth path, the progress in policy reforms has been frustratingly slow. There is a great risk that unless efforts are made to jumpstart their economies, the Central Asian countries may slide into a situation of "permanent transition," the social and political costs of which could be enormous.

One of the many reasons for this state is the lack of adequate information on the effect of alternative policy options facing the decisionmakers. Also, because external institutions instruct the governments of these countries on how to reform their economies and what kind of policy packages the reform program should contain, the governments do not develop the reforms themselves and hence lack ownership. Furthermore, externally imposed reforms may be inappropriate for the national circumstances. This problem is further compounded by several factors. First, institutional and human capacity for policy formulation and implementation has not been fully developed in the short period of less than a decade since independence. Second, external advisers seem to have largely ignored political economy issues in the design and execution of the policy reform packages. Third, social unrest, in the form of Islamic fundamentalism that may spread from Afghanistan to Tajikistan and other Central Asian republics, has forced the governments to take a cautious approach to policy reforms and to avoid any unintended negative impacts. Fourth, policy reforms have lost their priority on the governments' development agenda: preventing further deterioration of living conditions following the collapse of the Soviet Union has emerged as the major focus of the countries in the region. And finally, the basic macroeconomic framework fundamental to the success of sectoral policy reforms is largely missing in some of the Central Asian countries.

It is generally agreed that rethinking the policy reform process is necessary for generating momentum to reorient the nature, approach, and sequence of policy reform packages. This is particularly so in the food, agriculture, and natural resources sectors. To start, the national capacity for analyzing the impact of policy alternatives must be strengthened, so that policy reforms and the speed with which they are implemented are consistent with the objectives and the social and political realities of the individual countries in the region. Involving the local policy research community in identifying critical issues and challenges; setting priorities among them for food, agriculture, and natural resource policy research and analysis; and implementing joint research studies are the best ways to build local capacity and to increase the ownership of policy design and implementation. However, the analytical capacity is not well developed, and solid policy information remains scarce in this region.

In response to this need, and at the request of the Central Asian institutions involved in the process of policy reform, the International Food Policy Research Institute (IFPRI) and the Tashkent State Agrarian University jointly organized a regional workshop on "Food, Agriculture, and Natural Resource Policy Research: Setting the Priorities." This workshop was held July 19-21, 1999, in Tashkent, Uzbekistan. About 35 policy researchers and decisionmakers from the five Central Asian countries (Kyrgyz Republic, Kazakhstan, Tajikistan, Turkmenistan, and Uzbekistan), two other international agricultural research centers, International Center for Agricultural Research in the Dry Areas (ICARDA) and International Center for Maize and Wheat Improvement (CIMMYT), and several donor organizations participated in the workshop.

The goals of the workshop were these:

- to develop a regional understanding and consensus on the problems and required actions to alleviate poverty, reduce food insecurity, and sustainably manage the use of natural resources in Central Asia;
- to identify the information gaps in the process of policy reform; and
- to develop a set of priorities for food, agriculture, and natural resource policy research in Central Asia.

Dr. Abduvokhid Juraev, Deputy Minister for Agriculture and Water Resources, officially opened the workshop. In his opening remarks he recognized the need for information-based policymaking in Uzbekistan and in Central Asia. On behalf of IFPRI, Dr. Per Pinstrup-Andersen welcomed the participants. He underlined the importance of the agricultural sector in promoting overall economic growth in Central Asia and emphasized the need for higher levels of investment in agricultural research to increase the productivity of food and agricultural commodities.

During the first session of the workshop, key overview papers were presented addressing various food, agriculture, and natural resource policy issues specific to the Central Asia region. Group discussions for identifying key research issues followed the overview presentations. The outcome of the group discussions was presented to all the participants at a plenary session.

Dr. Rasulmat Khusanov, Director of the Market Reform Institute, Tashkent, delivered the introductory paper on the policy reform process in Uzbekistan. In his presentation, Dr. Khusanov emphasized the need for increasing the opportunities for regional trade among the five countries. Identifying various constraints to its agricultural development, he suggested reorganizing the food-processing sector in Uzbekistan. In the following presentations, Rajul Pandya-Lorch provided an overview of prospects for global food security to 2020 with a focus on

Central Asia's food demand, supply, and trade prospects and identified various issues for improving food security of the region; Suresh Babu and Alisher Tashmatov presented a set of emerging challenges for policy research for the Central Asia region; and Franscesco Goletti presented an overview of issues related to the functioning of input and output markets. In the final overview presentation, Per Pinstrup-Andersen, in his paper on linking the policy research needs of the Central Asian region to the research mandates of IFPRI, emphasized the need for undertaking research based on field surveys conducted in collaboration with national institutions.

Country teams, consisting of the participants from each of the five countries, prepared and presented papers addressing the key food, agriculture, and natural resource policy issues faced by their respective countries. Alisher Tashmatov, Roman Ospanov, and Eleonora Gaziyants presented the Uzbekistan paper; Kachkinbay Kadyrkulov presented the Kyrgzystan paper; Bakhriddin Amirov and Rakhmatullo Ergashev presented the Tajikistan paper; Adilya Baydildina presented the Kazakhstan paper; and Anadjemal Halnepesova presented the Turkmenistan paper. In addition to these country papers, three presentations addressing specific issues were also made during this session. Peter Oram discussed policy research needs for the rangelands and livestock systems in Central Asia. His paper was complemented by a presentation by Mekhlis Suleimenov of ICARDA on the trends in the feed and livestock production during the transition period in Kazakhstan, Kyrgyz Republic, and Uzbekistan. Erika Meng of CIMMYT made a presentation on the wheat economy of Kazakhstan.

This volume provides a compilation of the papers presented at the workshop. The country-level papers originally prepared in Russian have been translated into English. The papers are presented as chapters in three parts. As part of the introduction, the following chapter by Rajul Pandya-Lorch discusses global food security issues in a Central Asian context. In Part II, six chapters address the research needs for policy reforms in Central Asia. The country-level chapters prepared by the participants from the five Central Asian republics are given in Part III. The regional and country-specific policy research priorities are summarized in Part IV. This volume also contains the list of participants and their addresses for future contacts and collaboration.

2 Prospects for Global Food Security: A Central Asian Context

RAJUL PANDYA-LORCH

Dramatic transformations have occurred in recent decades in where and how food is produced, processed, and traded such that enough food is now available to meet the basic needs of each and every person in the world. The doubling of grain production and tripling of livestock production since the early 1960s has resulted in about 2,700 calories available per person per day. However, about 820 million people lack access to sufficient food to lead healthy and productive lives and around 160 million children are seriously underweight for their age. At the close of the 20th century, astonishing advances in agricultural productivity and human ingenuity have not yet translated into a world free of hunger and malnutrition.

What are the prospects for global food security in the 21st century? And what is the outlook for the Central Asia region, comprising Kazakhstan, Kyrgyz Republic, Tajikistan, Turkmenistan, and Uzbekistan, and covering 3,888 square kilometers with a total population of 56 million. This region has undergone massive transformations during the past decade following the dissolution of the Soviet Union and the transition from centrally planned economies to market-oriented economies. What are the key emerging issues in global food security that are likely to influence food security in Central Asia in coming decades?

Projections of food production and consumption to the year 2020 offer some signs of progress. But prospects of a food-secure world in which each and every person is assured of access at all times to the food required to lead a healthy and productive life remain bleak if the global community continues with business as usual. The International Food Policy Research Institute (IFPRI) has developed a global economic model, the International Model for Policy Analysis of Commodities and Trade (IMPACT),[1] which projects that under the most likely, or baseline, scenario about 133 million children under the age of six years will be malnourished in 2020, just 17 percent fewer than in 1993 (Table 2.1).[2] One out

---

[1] IMPACT is a global food model that divides the world into 37 countries and regions (which account for virtually all of the world's food production and consumption). It covers 18 commodities (including all cereals, soybeans, roots and tubers, meats, and dairy products) (Rosegrant, Agcaoili-Sombilla, and Perez 1995; Rosegrant, Ringler, and Gerpacio 1997).

[2] Malnourished children are those whose weight-for-age is more than two standard deviations below the weight-for-age standard set by the U.S. National Center for Health Statistics and adopted by many United Nations agencies in assessing the nutritional status of persons in developing countries.

TABLE 2.1 Number and percentage of malnourished children, 1995 and 2020

	Number		Percentage	
Region	1995	2020	1995	2020
	(millions)			
Latin America	4.9	2.6	9.1	4.8
West Asia and North Africa	5.6	5.0	13.2	10.5
China	18.0	12.6	17.4	13.7
Southeast Asia	18.7	14.1	33.7	26.8
Sub-Saharan Africa	31.9	39.2	32.9	28.6
South Asia	82.0	59.6	50.9	38.4
Developing world	161.1	133.1	31.4	24.7

SOURCE: IFPRI IMPACT simulations (March 1999).

of every four children in developing countries will be malnourished in 2020, down from 31 percent in 1995. Child malnutrition is expected to decline in all major developing regions except Sub-Saharan Africa, where the number of malnourished children could increase by more than 20 percent between 1995 and 2020 to reach 39 million. In South Asia, home to half of the world's malnourished children in 1995, the number of malnourished children is projected to decline by more than 20 million between 1995 and 2020. Still, the incidence of malnutrition is so high that, even with this reduction, two out of five children could remain malnourished in 2020 (Table 2.1). With about 75 percent of the world's malnourished children, Sub-Saharan Africa and South Asia are expected to remain hot spots of child malnutrition in 2020.

Projections by the Food and Agriculture Organization of the United Nations (FAO) on the number of chronically undernourished or food-insecure people paint a similarly mixed picture. FAO projects that 680 million people, 12 percent of the developing world's population, could be food insecure in 2010, down from 827 million in 1994-96 (Table 2.2). Food insecurity is expected to diminish rapidly in East Asia and, to a lesser extent, in South Asia and Latin America, but it could accelerate substantially in Sub-Saharan Africa and West Asia and North Africa. Sub-Saharan Africa and South Asia, home to a projected 70 percent of the world's food-insecure people in 2010, will be the locus of hunger in the developing world. In fact, Sub-Saharan Africa's share of the world's food-insecure population is projected to almost quadruple between 1969-71 and 2010 from 11 to 39 percent (FAO 1996). By 2010, every 3rd person in Sub-Saharan Africa is likely to be food insecure compared with every 8th person in South Asia and every 20th person in East Asia. These disturbing figures reflect widespread poverty and poor health.

TABLE 2.2 Number of food-insecure people, 1994-96 and 2010 (millions)

	1994-96	2010
East Asia	258	123
South Asia	254	200
Sub-Saharan Africa	210	264
Latin America	63	40
West Asia and North Africa	42	53
Total	827	680

SOURCE: FAO (1998).

Worldwide, per capita availability of food is projected to increase around 7 percent between 1995 and 2020, from about 2,700 calories to around 2,900 calories per day. Increases in average per capita food availability are expected in all major regions, including Central Asia and the rest of the former Soviet Union, which would experience increases of 6.1 percent and 4.5 percent respectively (Table 2.3).

Demand for food is influenced by a number of forces, including population growth and movements, income levels and economic growth, human resource development, and lifestyles and preferences. Under the medium-variant scenario, almost 80 million people are likely to be added to the world's population each year during the next two decades, increasing world population by 35 percent from 5.69 billion in 1995 to 7.67 billion by 2020 (UN 1996). More than 95 percent of the population increase is expected in developing countries, whose share of global population is projected to increase from 79 percent in 1995 to 84 percent in 2020. Over this period, the absolute population increase will be highest in Asia (1.15 billion), but the relative increase will be greatest in Sub-Saharan Africa, where the population is expected to double by 2020. Central Asia's population is expected to increase from 54 million in 1995 to 75 million in 2020, with more than half of the increase occurring in Uzbekistan. The population of the rest of the former Soviet Union is expected to decrease from 235 million in 1995 to 218 million in 2020, primarily because of substantial reductions in Russia and Ukraine. Within Central Asia, the population growth rate is expected to rise in Kazakhstan and Kyrgyzstan, but to decline in Tajikistan, Turkmenistan, and Uzbekistan (Table 2.4).

People's access to food depends on income. Currently, more than 1.3 billion people are absolutely poor, with incomes of a dollar a day or less per person, while another 2 billion people are only marginally better off with incomes of up to $2 a day (World Bank 1997). Within Central Asia, about 2 percent of Kazakhstan's population, 19 percent of Kyrgyzstan's population, and 5 percent of Turkmenistan's population have estimated incomes of less than $1 a day, while

TABLE 2.3 Per capita calorie availability, 1995 and 2020 (calories per day)

Region	1995	2020
Sub-Saharan Africa	2,144	2,295
South Asia	2,357	2,652
Central Asia	2,685	2,850
Southeast Asia	2,622	2,876
Rest of former Soviet Union	2,801	2,928
Latin America	2,789	3,026
China	2,752	3,139
West Asia and North Africa	3,081	3,177
Eastern Europe	3,055	3,206
EC-15	3,367	3,555
Developed countries	3,185	3,352
Developing countries	2,579	2,821
World	2,717	2,918

SOURCE: IFPRI IMPACT simulations (March 1999).

12 percent, 55 percent, and 26 percent, respectively, have incomes of less than $2 a day (World Bank 1999). These estimates are from 1993; undoubtedly, the situation is quite different today, with anecdotal evidence suggesting that poverty has deepened in this region.

Income growth rates have varied considerably in recent years, with Sub-Saharan Africa and West Asia and North Africa struggling with low or even negative growth rates while East Asia until the last two years was experiencing annual

TABLE 2.4 Medium-variant population projections for Central Asia, 1995-2020

	Level		Annual rate of change	
Country	1995	2020	1995-2000	2015-2020
	(millions)		(percent)	
Kazakhstan	16.8	18.7	0.13	0.73
Kyrgyzstan	4.5	5.3	0.37	1.17
Tajikistan	5.8	8.5	1.87	1.54
Turkmenistan	4.1	5.7	1.89	1.21
Uzbekistan	22.8	32.3	1.89	1.25

SOURCE: UNDP (1996).

growth rates exceeding 7 percent. Each of the five countries of Central Asia experienced negative growth during the 1990s, with growth rates for gross domestic product (GDP) during 1990-97 ranging from –3.5 percent in Uzbekistan to –9.6 percent in Turkmenistan, –10.5 percent in Kazakhstan, –12.3 percent in Kyrgyzstan, and –16.4 percent in Tajikistan. Prospects for economic growth during the next quarter century appear favorable. At 3 percent, the average annual income growth rate projected by IMPACT for Central Asia during 1995-2020 is higher than the rate of 2 percent projected for the rest of the former Soviet Union and lower than 4 percent projected for Eastern Europe. Growth rates are projected to be higher in South and Southeast Asia, including China.

Under the baseline scenario, IMPACT projects global demand for cereals to increase by 39 percent between 1995 and 2020 to reach 2,476 million metric tons, for meat demand to increase by 59 percent to 314 million tons, and for roots and tubers demand to increase by 37 percent to 868 million tons. Developing countries will drive increases in world food demand. With a projected 40-percent population increase and an average annual income growth rate of 4.3 percent, developing countries are expected to account for most of the increase in global demand for cereals and meat between 1995 and 2020 (Table 2.5). Central Asia is forecast to account for less than 1 percent of the increase in global demand for cereals and meat between 1995 and 2020. Note that while Central Asia is projected to experience a 33-percent increase in its demand for cereals, the rest of the former Soviet Union is expected to experience virtually no increase in its demand for cereals.

TABLE 2.5 Global demand for cereals and meat, 1995 and 2020 (millions of tons)

	Cereal demand			Meat demand		
Region	1995	2020	Increase 1995-2020	1995	2020	Increase 1995-2020
Central Asia	18.06	24.01	5.95	1.95	2.86	0.91
Rest of former Soviet Union	115.74	116.58	0.84	11.53	12.34	0.81
West Asia and North Africa	120.20	191.95	71.75	6.91	13.38	6.47
Sub-Saharan Africa	77.60	153.16	75.56	5.06	10.94	5.88
Latin America	137.35	217.72	80.37	23.04	42.39	19.35
South Asia	226.17	355.99	129.82	6.86	15.18	8.32
East and Southeast Asia	511.25	743.64	232.39	57.56	147.74	90.18
Developed countries	704.49	812.48	107.99	98.33	116.50	18.17
Developing countries	1,073.26	1,663.65	590.39	99.63	197.67	98.04
Total	1,777.75	2,476.14	698.39	197.96	314.17	116.21

SOURCE: IFPRI IMPACT simulations (March 1999).

Between 1995 and 2020, per capita demand for cereals is projected to increase by 10 kilograms in Central Asia, by 29 kilograms in the rest of the former Soviet Union, and by 62 kilograms in Eastern Europe, while per capita demand for meat is projected to increase by 5, 6, and 15 kilograms, respectively (Table 2.6). Disparities in consumption are expected to remain wide between the developed and developing world: a developing-country person is forecast to consume, on average, only 45 percent of the cereals, 37 percent of the meat, and 71 percent of the roots and tubers that a developed country person would consume in 2020.

Demand for cereals for feeding livestock will increase considerably in importance in coming decades, especially in developing countries, in response to strong demand for livestock products. Between 1995 and 2020, global demand for cereals for animal feed is projected to increase by 45 percent, while demand for cereals for direct human consumption is projected to increase 36 percent. By 2020, 38 percent of the cereal demand will be for feed, compared with 36 percent in 1995. In Central Asia, demand for cereals for feed is projected to increase from 3.88 million tons to 5.31 million tons between 1995 and 2020, and demand for cereals for direct human consumption is projected to increase from 10.69 million tons to 14.06 million tons. In the rest of the former Soviet Union, demand for cereals for direct human consumption is projected to decline from 34.76 million tons to 31.39 million tons, but demand for feed is projected to increase from 59.29 million tons to 63.28 million tons.

TABLE 2.6 Per capita demand for cereals and meat, 1995 and 2020 (kilograms)

	Cereal demand			Meat demand		
Region	1995	2020	Increase 1995-2020	1995	2020	Increase 1995-2020
Central Asia	335	345	10	36	41	5
Rest of former Soviet Union	484	513	29	48	54	6
Eastern Europe	629	691	62	67	82	15
West Asia and North Africa	370	380	10	21	27	6
Sub-Saharan Africa	142	156	14	9	11	2
Latin America	292	334	42	49	65	16
South Asia	182	200	18	6	9	3
East Asia	307	376	69	38	63	25
Southeast Asia	238	251	13	17	27	10
Developed countries	547	597	50	76	86	10
Developing countries	246	272	26	23	32	9
Total	315	331	16	35	42	7

SOURCE: IFPRI IMPACT simulations (March 1999).

The expected increases in cereal demand around the world will have to be met primarily by increases in productivity. Increases in cultivated area will contribute less than 20 percent of the increase in global cereal production between 1995 and 2020. Most of the growth in cereal area will be concentrated in the relatively low productivity cereals in Sub-Saharan Africa. There will be some expansion in Latin America but cereal area will remain virtually stagnant in Asia. In Central Asia, area under cereals is projected to increase at an average annual rate of growth of 0.25 percent between 1995 and 2020, a much higher rate of increase than the 0.03 percent growth rate projected for the rest of the former Soviet Union (Table 2.7).

Since growth in cultivated area is unlikely to contribute much to future production growth, the burden of meeting increased demand for cereals rests on improvements in crop yields. However, the annual increase in yields of the major cereals is projected to slow down during 1995-2020 compared with preceding decades in both developed and developing countries. This is worrisome given that yield growth rates were already on the decline. The two key reasons for slower cereal yield growth rates are as follows: (1) in regions where input use is high, such as Asia, farmers are approaching economically optimum yield levels, making it more difficult to sustain the same rates of yield gains; and (2) declining world cereal prices are causing farmers to switch from cereals to other, more profitable, crops and are causing governments to slow their investment in agricultural research and irrigation and other infrastructure.

With the projected slowdowns in area expansion and yield growth, cereal production in developing countries as a group is also forecast to slow to an annu-

TABLE 2.7 Annual growth rates in cereal production, 1995-2020 (percent)

Region	Area	Yield	Production
Central Asia	0.25	0.88	1.13
Rest of former Soviet Union	0.03	0.56	0.59
Eastern Europe	0.10	1.10	1.21
West Asia and North Africa	0.35	1.47	1.82
Sub-Saharan Africa	1.18	1.58	2.78
Latin America	0.55	1.54	2.10
South Asia	0.17	1.43	1.60
East Asia	0.11	1.19	1.30
Southeast Asia	0.18	1.33	1.51
Developed countries	0.08	0.83	0.91
Developing countries	0.41	1.25	1.66
Total	0.29	1.04	1.33

SOURCE: IFPRI IMPACT simulations (March 1999).

al rate of 1.7 percent during 1995-2020 compared with 2.3 percent during 1982-94. This figure is still higher, however, than the 0.9 percent annual rate of growth projected for developed countries during 1995-2020.

IMPACT projections suggest that food production will not keep pace with demand in developing countries and an increasing portion of the developing world's food consumption will have to be met by imports from the developed world. The proportion of cereal demand that is met through net imports is projected to rise from 10 percent in 1995 to 12 percent in 2020. As a group, developing countries are projected to more than double their net imports of cereals (the difference between demand and production) between 1995 and 2020 (Table 2.8). With the exception of Latin America, all major developing regions are projected to significantly increase their net cereal imports: the 166 percent increase in Asia's net imports will be driven primarily by rapid income growth, while the doubling forecast for Sub-Saharan Africa will be driven primarily by its continued poor performance in food production and rapid population growth. Central Asia is expected to remain a minor net importer of cereals, while the rest of the former Soviet Union is expected to switch from being a net importer of cereals to a substantial net exporter, providing about 6 percent of the developed world's exports to developing countries by 2020. The European Union is forecast to provide 15 percent of the net cereal imports of developing countries in 2020, the United States about 60 percent, and Australia about 10 percent.

With continued population growth, rapid income growth, and changes in lifestyles, demand for meat is projected to increase by 2.8 percent per year dur-

TABLE 2.8 Net trade in cereals, 1995 and 2020 (millions on tons)

Region	1995	2020
Central Asia	–0.51	–0.76
Rest of former Soviet Union	–5.27	+11.31
Eastern Europe	+0.34	+21.40
West Asia and North Africa	–37.81	–62.50
Sub-Saharan Africa	–9.83	–18.69
Latin America	–20.37	–21.05
South Asia	–0.27	–19.92
East Asia	–31.07	–73.97
Southeast Asia	–7.03	–8.19
Developed countries	+107.03	+205.50
Developing countries	–107.03	–205.50

SOURCE: IFPRI IMPACT simulations (March 1999).

NOTE: Minus sign indicates net imports; plus sign indicates net exports.

ing 1995-2020 in developing countries and by 0.7 percent per year in developed countries. While per capita demand for cereals in developing countries is projected to increase by only 10 percent between 1995 and 2020, per capita demand for meat will increase by more than 40 percent. The increase in per capita meat demand will be largest in East Asia and smallest in Sub-Saharan Africa and South Asia; by 2020, Chinese per capita consumption of meat will be more than eight times that of India (Table 2.6). Meat production is expected to grow by 2.7 percent per year in developing countries during 1995-2020 (compared with 5.9 percent during 1982-94) and by 0.8 percent in developed countries (compared with 0.9 percent during 1982-94). Despite high rates of production growth, developing countries as a group are projected to increase their net meat imports sixfold, reaching 5.1 million tons in 2020 (Table 2.9). Latin America will continue to be a net exporter of meat, but Asia will switch from being a small net exporter to a large net importer. Central Asia is projected to more than double its net meat imports between 1995 and 2020 to 0.38 million tons, but the rest of the former Soviet Union is projected to experience a slight reduction in its net meat imports to 1.62 million tons.

Net imports are a reflection of the gap between production and market demand. For many of the poor, the gap between food production and human needs is likely to be even wider than that between production and demand because many of these people are priced out of the market, even at low food prices, and are unable to exercise their demand for needed food. The higher

TABLE 2.9 Net trade in meat, 1995 and 2020 (million of tons)

Region	1995	2020
Central Asia	–0.16	–0.38
Rest of former Soviet Union	–1.85	–1.62
Eastern Europe	+0.06	–0.20
West Asia and North Africa	–0.98	–1.82
Sub-Saharan Africa	–0.15	–0.14
Latin America	+0.53	+1.78
South Asia	+0.14	–0.34
East Asia	–0.19	–3.28
Southeast Asia	–0.04	–0.95
Developed countries	+0.81	+5.07
Developing countries	–0.81	–5.07

SOURCE: IFPRI IMPACT simulations (March 1999).

NOTE: Minus sign indicates net imports; plus sign indicates net exports.

income developing countries, notably those of East Asia, will be able to fill the gap between production and demand through commercial imports, but the poorer countries may be forced to allocate foreign exchange to other uses and thus might not be able to import food in needed quantities. It is the latter group of countries, including most of those in Sub-Saharan Africa and some in Asia, which will remain a challenge and require special assistance to avert widespread hunger and malnutrition.

Emerging Issues in Global Food Security of Relevance to Central Asia

Volatile Cereal Prices

Enough food will be produced to meet the demand of people who can afford to buy it such that world food prices will continue to decline over the next two decades, although at slower rates than in the past (Table 2.10). However, concerns are growing that cereal prices may be more volatile than in the past. Reduced stocks and uncertainties associated with developments in China and the former Soviet Union, among other factors, could increase price instability. On the other hand, market liberalization in developing countries, policy reform in developed countries, and more consistent and transparent stock-holding and trade policies will make producers more responsive to price changes and could reduce price instability. How these factors play out will determine whether cereal prices will be more volatile in coming years. In addition to price fluctuations in the international market, many countries suffer from large domestic or local price fluctuations owing to inadequate markets, poor roads and other infrastructure, and inappropriate policies and institutions. Even small changes in production resulting from better or poorer growing conditions may cause large fluctuations in local food prices.

Growing Water Scarcity

Unless properly managed, fresh water may well emerge as the key constraint to global food production. While supplies of water are adequate in the aggregate to meet demand for the foreseeable future, water is poorly distributed across countries, within countries, and between seasons. And, with a fixed amount of renewable water resources to meet the needs of a continually increasing population, per capita water availability is declining steadily. Demand for water is growing rapidly. Since 1970, global demand for water has grown by 2.4 percent per year (Rosegrant, Ringler, and Gerpacio 1997). Projections of water demand to 2020 indicate that global water withdrawals will increase by 35 percent between 1995 and 2020 to reach 5,060 billion cubic meters. Developed countries are projected to increase their water withdrawals by 22 percent, more than 80 percent of the

TABLE 2.10 World food prices, 1995-2020

Year	Beef	Rice	Maize	Wheat
	(US$ per 100 kilograms)	(US$ per ton)		
1995	202.8	336	116	162
1996	202.2	338	116	161
1997	202.0	341	116	162
1998	201.8	344	117	162
1999	201.5	346	117	162
2000	201.1	349	117	162
2001	201.0	350	118	162
2002	200.7	351	118	162
2003	200.3	351	118	161
2004	199.8	351	118	161
2005	199.4	351	119	160
2006	199.0	350	119	160
2007	198.6	348	119	159
2008	198.1	346	119	158
2009	197.5	343	118	157
2010	196.9	340	118	156
2011	196.4	337	118	155
2012	195.8	333	118	154
2013	195.1	329	118	152
2014	194.4	324	118	151
2015	193.6	320	117	150
2016	192.9	315	117	148
2017	192.2	309	117	146
2018	191.4	304	116	145
2019	190.4	298	116	143
2020	189.5	292	115	141

SOURCE: IFPRI IMPACT simulations (March 1999).

increase being for industrial uses. Developing countries are projected to increase their withdrawals by 43 percent over the same period and to experience a significant structural change in their demand for water, reducing the share for agricultural uses.

The costs of developing new sources of water are high and rising, and nontraditional sources such as desalination, reuse of wastewater, and water harvesting are unlikely to add much to global water availability, although they may be important in some local or regional ecosystems. So how can the rapid increases in water demand be met? The rapidly growing domestic and industrial demand for water will have to be met from reduced use in the agriculture sector, which is by far the largest water user, accounting for 72 percent of global water withdrawals and 87 percent of withdrawals in developing countries in 1995 (Rosegrant, Ringler, and Gerpacio 1997). Reforming policies that have contributed to the wasteful use of water offers considerable opportunity to save water, improve efficiency of water use, and boost crop output per unit of water. Required policy reforms include establishing secure water rights for users; decentralizing and privatizing water management functions; and setting incentives for water conservation, including markets in tradable water rights, pricing reform and reduction in subsidies, and effluent or pollution charges (Rosegrant 1997). Failure to address the gap between tightening supplies and increasing demand for water could significantly slow growth in food production.

Declining Soil Fertility

Improved soil fertility is a critical component of the drive to increase sustainable agricultural production. Past and current failures to replenish soil nutrients in many countries must be rectified through the balanced and efficient use of organic and inorganic plant nutrients and through improved soil management practices. Although some of the plant nutrient requirements can be met through the application of organic materials available on the farm or in the community, such materials are insufficient to replenish the plant nutrients removed from the soils and thus to further expand crop yields. But the use of chemical fertilizers has decreased worldwide during the last few years, particularly in the developed countries and in parts of Asia. Although reduced use of fertilizers is warranted in some locations because of negative environmental effects, it is critical that fertilizer use be expanded in countries where soil fertility is low and a large share of the population is food insecure. Fertilizer consumption in these countries is generally low because of high prices, insecure supplies, and the greater risk associated with food production in marginal areas.

In view of the size and seriousness of the soil fertility problem in many countries, a cost-effective fertilizer sector and policies providing incentives for farmers and communities to implement soil fertility programs are needed. Such policies should focus on supporting agricultural research to generate (1) appropriate technology; (2) clear long-term property rights to land; (3) access to credit, improved crop varieties, water, and information about effective and efficient fertilizer use in various production systems; (4) efficient and effective markets for plant nutrients; and (5) investments in roads and rural transportation systems. Of particular importance to maintaining and enhancing soil fertility is the adoption of integrated plant nutrient management (IPNM) practices. The goal of IPNM is to integrate the use of natural and human-made sources of plant nutrients to increase agricultural productivity in an efficient and environmentally benign manner without diminishing the productive capacity of soil for present and future generations.

Escalating Concerns about Food Safety

Concerns about food safety are not new. Since time immemorial, human beings have worried about whether they have sufficient food to eat and whether the food they consume is safe and healthy. However, food safety concerns are escalating, particularly in industrialized countries, as evidenced by the growing demand for organic foods; the strengthening public backlash against genetically modified foods; the extraordinarily high level of consumer interest in the precise origin and modes of producing and processing the food they consume; and the proliferation of regulations on producing, processing, storing, and transporting foods. There have been a series of well-publicized outbreaks of food-borne illnesses and massive food recalls in recent years, particularly in the United States and Western Europe.

The growing concerns over food safety have two major implications. First, exports of food commodities could be exposed to new and more demanding food safety standards, partly through changes in the Codex Alimentarius and partly through unilateral demands by importers. (Thus, food safety requirements could become a hindrance to realizing benefits from exports if either the exporting countries cannot meet unreasonable standards or the importing countries use food safety as a nontariff barrier.) Second, changing attitudes toward legislation for food safety in industrialized countries could spill over into other countries, particularly developing countries, without due attention to local conditions, constraints, and influence such as availability of and access to food. (For example, legislation to curtail or prohibit the use of fertilizers or chemical plant protection methods could have a negative effect on food security through increased unit costs of productions.)

Weather Fluctuations and Climate Change

With the recent resurgence of El Niño, followed by the ongoing La Niña, major weather fluctuations are under way or imminent in many parts of the world. These weather fluctuations could lead to sizable food production shortfalls and deterioration in food security in many parts of the world. The 1997/98 El Niño caused severe drought in Southeast Asia, flooding in the Andean countries of South America, and drought in a wide swath across Eastern Africa. El Niño adds a major element of uncertainty to agricultural production and livelihoods around the world. And concerns are growing that El Niño may become more frequent and more severe in the future as a result of climate changes.

Although the trend of global warming is becoming increasingly clear, its effects on food production are still uncertain. Some research suggests that growing conditions will deteriorate in current tropical areas (where many of the developing countries are located) and improve in current temperate areas (where many of the developed countries are located) (Rosenzweig and Parry 1994; Fischer et al. 1996). However, effects on productivity and production will occur over a long period of time and will be very small in any given year. Therefore, it is reasonable to believe that policies and technologies can be developed to effectively prevent or counter the negative productivity effects of global warming.

Conflicts and Food Security

Widespread local, national, and regional instability and armed conflicts contribute to the persistence of poverty, food insecurity, and natural resource degradation. While relief agencies around the world are fully aware of the disastrous effects of conflicts on peoples' capacity to ensure their food security, opportunities for preventing or resolving conflict through improvements in food security and more sustainable use of natural resources have received little attention until recently. It is becoming increasingly clear that poverty, food insecurity, and natural resource degradation contribute to the initiation or prolongation of instability or conflicts. Poor, food-insecure people may, in desperate circumstances, perceive no option but to engage in conflict to secure their access to resources that will assure future well-being. Of course, not all poor, food-insecure people engage in conflict, but the probability of instability or conflict rises in circumstances where people are pushed to the limit to meet even their most fundamental needs. The complex, mutually reinforcing relationship between poverty, food insecurity, and natural resource degradation on the one hand, and social and political instability and conflict on the other hand, has not been fully recognized and acknowledged.

Trade Liberalization Issues

In response to the recent General Agreement on Tariffs and Trade (GATT) agreement and structural adjustment programs, a large number of countries have liberalized foreign trade in food and agricultural commodities. Unfortunately, this has not been matched by market openings in the countries of the Organization for Economic Cooperation and Development (OECD). While preferential treatments are still in place for specific quantities of certain commodities, OECD countries have been reluctant to open up their domestic markets for imports of high-value commodities such as beef, sugar, groundnuts, and dairy products. At the same time, developing countries are being encouraged through structural adjustment to emphasize production of these same high-value agricultural commodities for export. From the point of view of food security and poverty alleviation in developing countries, the next round of World Trade Organization (WTO) negotiations should emphasize the opening up of domestic markets in OECD countries for commodities from developing countries.

To fully benefit from trade liberalization, developing countries must invest in domestic infrastructure, effective and efficient agricultural input and output markets, research and technology, and a more equitable distribution of land and other productive resources. Furthermore, while most poor and food-insecure people are expected to benefit from trade liberalization, the distribution of benefits will be determined largely by the distribution of productive assets. Countries with very skewed distribution of assets may experience an amplification of this pattern as a result of trade liberalization. Therefore, emphasis on trade liberalization should be matched with similar or stronger emphasis on rectifying domestic policies, including improved access by the poor to productive assets and employment.

The breakup of the former Soviet Union has opened up the economies of these countries, but the pace of their transition to free market rules has varied greatly (Grote and Wehrheim 1999). Most of these countries of the Commonwealth of Independent States (CIS) are negotiating to join the WTO. Will WTO accession help them overcome agricultural growth problems? How relevant is WTO, given the domestic institutional constraints that conspire against better-functioning food markets? Under what conditions will their national agricultural trade systems be integrated into the WTO? Because of pressure from international donors, most quantitative export and import restrictions have been eliminated and export taxes have been reduced considerably in recent years. However, the issue of market access remains controversial. Substantial shares of the food trade, both within and outside the CIS, are still based on barter arrangements and state trading, resulting in discrimination against international, non-CIS competitors.

Accession of the CIS to WTO could stimulate domestic reforms either through mandatory WTO rules or as the result of negotiations with other WTO members. Domestic trade barriers within regions, such as those that exist in the CIS, and contract insecurity need to be overcome. Capacity building and institutional strengthening are needed to streamline agricultural policies and strategies. Domestic and regional market reforms in the CIS are a precondition for reaping WTO trade benefits. The negotiations for WTO accession should be intensified to speed up this process. A quick WTO accession could prevent any trend toward increasing protectionism. However, WTO accession will only result in more liberal agricultural trade regimes in the CIS if the millennium round trade negotiations promise further reduction of agricultural protection in the European Union (EU) and the United States. Only under these conditions will the transition countries gain better market access and thereby increase export opportunities and improve the economic stabilization process.

Conclusion

Food insecurity has long been perceived by some to be primarily a problem of insufficient food production rather than insufficient access to food. Yet, as enough food is being produced globally to meet the basic needs of every person in the world, it is evident that the persistence of food insecurity—about 820 million chronically undernourished people and 160 million malnourished children—is increasingly attributable to difficulties in accessing sufficient food. Food-insecure people simply do not have the means to grow and/or purchase the needed food. Ensuring that every individual has access to remunerative employment; to productive assets such as land and capital; and to productivity-enhancing resources such as appropriate technology, credit, education, and health care is essential. Besides enabling every person to acquire the means to grow and/or purchase sufficient food to lead healthy and productive lives, ensuring a food-secure world means producing enough food to meet increasing and changing food needs and meeting food needs through better management of natural resources.

With foresight and decisive action, we can create the conditions that ensure food security for all people in coming years. The action required is not new or unknown. Ensuring a food-secure world will require all relevant parties—individuals, households, farmers, local communities, the private sector, civil society, national governments, and the international community—to work together in new or strengthened partnerships. It will require a change in behavior, priorities, and policies. And it will require strengthened cooperation between developing and industrialized countries and among developing countries. The world's natural resources are capable of supporting sustainable food security for all people if current rates of degradation are reduced and replaced by appropriate technological change and sustainable use of natural resources.

We have the means to ensure a food-secure world; let us act to make it a reality for each and every person.

References

FAO (Food and Agriculture Organization of the United Nations). 1996. Investment in agriculture: Evolution and prospects. World Food Summit Technical Background Document 10. Rome.

_____. 1998. The state of food and agriculture 1998. Rome.

Fischer, G., K. Frohberg, M. L. Parry, and C. Rosenzweig. 1996. Impacts of potential climate change on global and regional food production and vulnerability. In Climate change and world food security, ed. T. E. Downing, NATO ASI Series, Vol. 37. Berlin: Springer Verlag.

Grote, U., and P. Wehrheim. 1999. Getting ready for the millennium round trade negotiations: Transition economies' perspective. 2020 Focus No. 1. Washington, D.C.: International Food Policy Research Institute.

Rosegrant, M. W. 1997. Water resources in the twenty-first century: Challenges and implications for action. Food, Agriculture, and the Environment Discussion Paper 20. Washington, D.C.: International Food Policy Research Institute.

Rosegrant, M. W., M. Agcaoili-Sombilla, and N. D. Perez. 1995. Global food projections to 2020: Implications for investment. Food, Agriculture, and the Environment Discussion Paper 5. Washington, D.C.: International Food Policy Research Institute.

Rosegrant, M. W., C. Ringler, and R. V. Gerpacio. 1997. Water and land resources and global food supply. Paper prepared for the 23rd International Conference of Agricultural Economists, Sacramento, California, 10-16 August.

Rosenzweig, C., and M. L. Parry. 1994. Potential impact of climate change on world food supply. Nature 367 (January 13): 133-138.

Rozelle, S., and M. W. Rosegrant. 1997. China's past, present, and future food economy: Can China continue to meet the challenges? Food Policy 22 (3): 191-200.

Tyers, R. 1994. Economic reform in Europe and the former Soviet Union: Implications for international food markets. Research Report 99. Washington, D.C.: International Food Policy Research Institute.

UN (United Nations). 1996. World population prospects: The 1996 revisions. New York: UN.

World Bank. 1997. 1997 World development indicators. Washington, D.C.

______. 1999. World development report. Washington, D.C.

PART II

Research Needs for Policy Reforms in Central Asia

3 IFPRI's Research Priorities and Linkages to Central Asia

PER PINSTRUP-ANDERSEN, STACY ROBERTS,
AND MARC J. COHEN

Imagine a world where (1) *every woman, man, and child has access to the food necessary for a healthy and productive life*; (2) *every child is well-nourished*; and (3) *people use natural resources sustainably and maintain biodiversity*. This is IFPRI's global vision for the year 2020 and the kind of world that IFPRI is trying to help achieve through our food policy research.

The world has made much progress toward achieving the vision. Just 30 years ago, annual per capita income in the developing world was about $700. Forty percent of the people were undernourished, and a majority lived in absolute poverty, on less than $1 a day. The average woman had six children during her lifetime. More than 50 percent of the people were illiterate, and life expectancy was around 50. Twelve million infants died annually, and three out of four people lacked access to clean water and safe sanitation.

Since then developing countries have improved by leaps and bounds. Malnutrition has fallen by more than 50 percent, per capita income has increased by 60 percent, the absolute poverty rate has fallen by almost half, life expectancy has increased by more than 10 years, literacy rates have increased by almost 50 percent, and infant mortality has fallen by 50 percent. Five million fewer children die each year. The average woman has three children during her lifetime.

Yet as the 21st century approaches, we remain far from the goal of eliminating hunger and malnutrition, even though the world produces enough food for every person on earth to enjoy an adequate diet. More than 800 million people (1 of every 5 people) in the developing world are still hungry. This is more than 13 times the population of Central Asia (Kazakhstan, Kyrgyzstan, Tajikistan, Turkmenistan, and Uzbekistan). The Asia-Pacific region is home to 512 million hungry people. The number of hungry people in South Asia, after decreasing in the 1980s, has increased by 17 million from 1992 to 1996. In Sub-Saharan Africa, 210 million or 39 percent of its people are malnourished. Food insecurity in Africa is expected to increase over the next decade (Figure 3.1).

As you know, four of the five Central Asian nations are considered to be low-income food deficit countries (LIFDCs) by the Food and Agriculture Organization of the United Nations (FAO). Kazakhstan is the exception. LIFDCs are particularly vulnerable to food insecurity because they have low levels of per capita income and are net importers of food in a time of volatile global prices. In Tajikistan, the lingering consequences of civil war have created additional problems for segments

FIGURE 3.1 Number of food-insecure people, 1979–81, 1994–96, and 2010

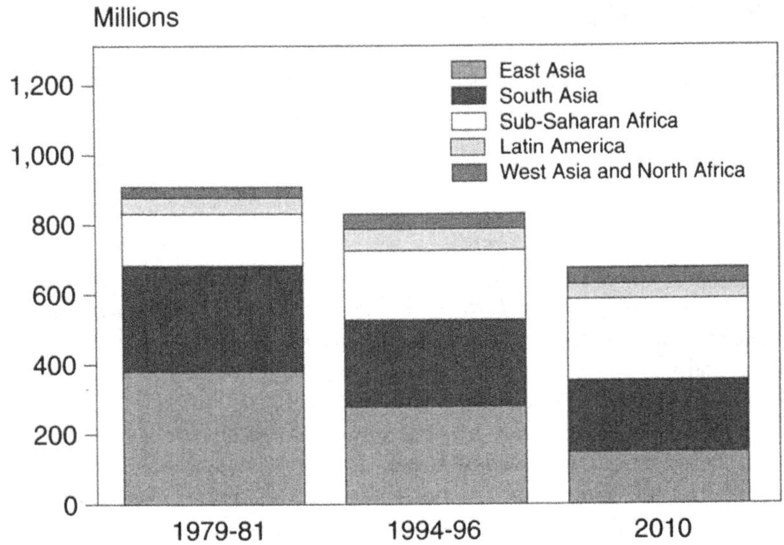

Source: FAO (1996 and 1998

of the population, especially those who remain internally displaced. As discussed in other papers prepared for this meeting, sustainable increases in agricultural productivity among small farmers in Central Asian and other LIFDCs are essential for both poverty reduction and long-term food security.

Close to 167 million children under age five in developing countries are malnourished. Malnutrition among preschool children can lead to permanent physical and mental damage, and is a factor in the deaths of more than 5 million children under age five each year. A majority of the malnourished preschoolers live in South Asia, where 49 percent of the children under age five suffer from malnutrition. According to our projections, if there are no significant changes in policy, child malnutrition in the developing world will decline by just 14 percent by 2020. Child malnutrition can be expected to rise 24 percent in Sub-Saharan Africa, although declines are forecast in all other developing regions (Figure 3.2).

About 70 percent of the world's poor people live in rural areas, and even if they do not engage directly in farming, nonfarm employment and incomes generally depend on agriculture. Moreover, agricultural growth is usually the most viable engine for overall broad-based economic growth and development in low-income countries. Very few countries have experienced rapid economic growth without agricultural growth either preceding it or accompanying it. While economic growth is not the sole ingredient needed to achieve poverty reduction and food security, poverty and food insecurity are unlikely to decrease without economic growth.

FIGURE 3.2 Number of malnourished children (0–5 years old), 1995, 2010, and 2020

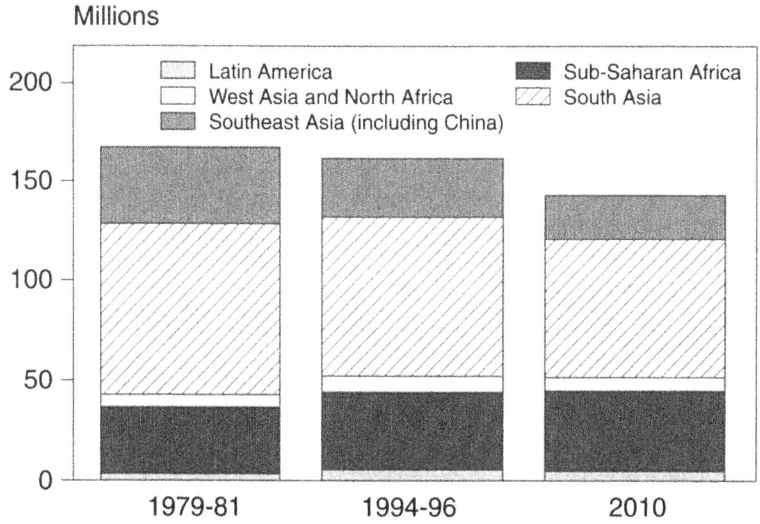

Source: WHO (1998): IFPRI IMPACT Projections

Agricultural research is essential to generate the technology needed by farmers to increase productivity. However, it alone will not drive agricultural growth. The interaction between technology and policy is critical. The full benefits of agricultural research and technological change will materialize only if government policies are appropriate. Moreover, achieving broad-based, equitable, and environmentally sustainable agricultural growth requires better utilization of the knowledge local farmers have gained over the past many years, as well as the use of all appropriate modern scientific tools, including applications of biotechnology oriented toward the problems facing poor farmers. Low-income farmers and consumers must participate actively in efforts to achieve food security. When the intended beneficiaries are treated as passive recipients and do not attain a sense of ownership, development efforts are not likely to succeed.

How does IFPRI work to achieve its vision of a hunger-free world by 2020? Our mission is to identify and analyze alternative national and international strategies and policies for meeting the food needs of the developing world on a sustainable basis, with particular emphasis on low-income countries, poor people, and sound management of the natural resource base that supports agriculture; to make the results of our research available to all those in a position to use them; and to help strengthen developing country institutions that conduct research and apply research results.

In particular, our research focuses on adding to the body of knowledge of food policy that can be used by many countries throughout the developing and developed world—what we call international public goods. In order to create this general knowledge, we collaborate with national institutions in on-the-ground studies to generate results that our research partners in specific study countries can utilize. Right now we are working with partners in more than 45 countries to create the knowledge needed to design and implement policies and programs that will advance sustainable food security. We share research results with key policy implementers and seek to strengthen the capacity of local institutions and individuals to conduct their own research on food policies. Our partner institutions include government agencies, universities, national agricultural research systems, and civil society organizations.

IFPRI is part of the global agricultural research network known as the Consultative Group on International Agricultural Research (CGIAR). Many of our activities involve collaboration with the other 15 research centers in this system, most of which carry out biological research.

IFPRI generates policy information through research and outreach activities in five overlapping areas. First, we study ways to improve markets for production inputs such as fertilizer and seeds as well as markets for farmers' crops. Related to this, we conduct research to help farmers diversify their agriculture so that they can spread their risks, boost their incomes, and function better in an increasingly global economy. Second, IFPRI seeks policies that increase food production and reduce poverty without degrading soil or water resources. Third, we assess various aspects of macroeconomic and trade policy that affect the production, supply, and price of food. Fourth, IFPRI conducts research that promotes food security and good nutrition for all, particularly for women and children. Finally, we seek to strengthen the effectiveness and efficiency of agricultural research and technology. Throughout our research work, we strive to disseminate food policy research outputs to targeted audiences and to strengthen the capacity of developing-country collaborators to undertake food policy research. Over the past three years, we have also devoted substantial attention to assessing the impact of our work.

To explain more clearly how IFPRI works, we will provide a few examples of projects from each of our research and outreach areas.

Institutions, Markets, and Infrastructure

Working with a number of units within the Ministry of Agriculture and Rural Development (MARD) of Viet Nam, IFPRI undertook an in-depth investigation of the rice marketing system in the midst of the country's transition from a centrally planned to a market-based economy. From a survey of approximately 1,300 rice farmers, millers, and traders, we found that the private sector had responded strongly to the reforms the government had instituted, yet a number of constraints remained that blocked further market expansion. Our study found that liberalizing external trade (quota removal) and internal trade would lead to a considerable income gain for the country while benefiting farmers and rural poor people. Over the course of the project, we also carried out a variety of outreach activities. For example, we offered short training courses in food policy analysis for officials of MARD and other government agencies, ran numerous workshops and seminars in Hanoi and the provinces, worked with the international media, and disseminated IFPRI publications. An independent consultant assessed the impact of the project and estimated that relaxing rice export quotas and internal restrictions on the rice trade in 1995-97 generated a gain of $61 million for the Vietnamese economy. This study found that IFPRI's research work satisfied policymakers that rice market liberalization would not adversely affect regional disparities and food security, and would have a beneficial effect on farm prices and poverty. The study also strengthened reform advocates within the policymaking process.

We think our research in Viet Nam may offer some valuable lessons for Central Asian governments as they continue to steer difficult transitions from centrally planned to market-oriented economies. For example, Kazakhstan, like Viet Nam, has historically produced grain for export. What policies would help to sustain and strengthen grain exports in a more market-oriented future?

An example of IFPRI's work to diversify income strategies for the rural poor is found in our research on livestock. A revolution is taking place in the demand for livestock products. Population growth, urbanization, and income growth in developing countries is fueling a massive increase in demand for food of animal origin. Research by IFPRI, another CGIAR center (the International Livestock Research Institute), and FAO predicts that consumption of meat and milk will grow 2.8 and 3.3 percent per year, respectively, between the early 1990s and 2020. In developed countries the rates will only grow 0.6 and 0.2 percent per year. Meat consumption will grow from 184 million metric tons in 1993 to 303 million metric tons in 2020. Demand for grain for use as animal feed will also grow in the developing world, driving increased cereal imports (Figure 3.3). By 2020 developing countries will produce 60 percent of the world's meat and 52 percent of the world's milk. The increased consumption of livestock products could significantly improve the health and nutrition levels of many rural poor people, and small-scale animal husbandry offers new income earning opportunities. Policies and investments need to be in place that will make the meat revolution as beneficial as possible for poor people. This research is particularly relevant for Kyrgyzstan, as it seeks to modernize the sheep subsector.

Agricultural Intensification

Within this area, one of our research programs focuses on water resource management policies, which are critical for Central Asia. Unless properly managed, fresh water may well emerge as the most important constraint to global food production over the next two decades. While supplies of water are adequate in the aggregate to meet the demand in the foreseeable future, water is poorly distributed across countries, within countries, and between seasons. By 2025, 50 coun-

FIGURE 3.3 Net cereal imports of major developing regions, 1993 and 2020

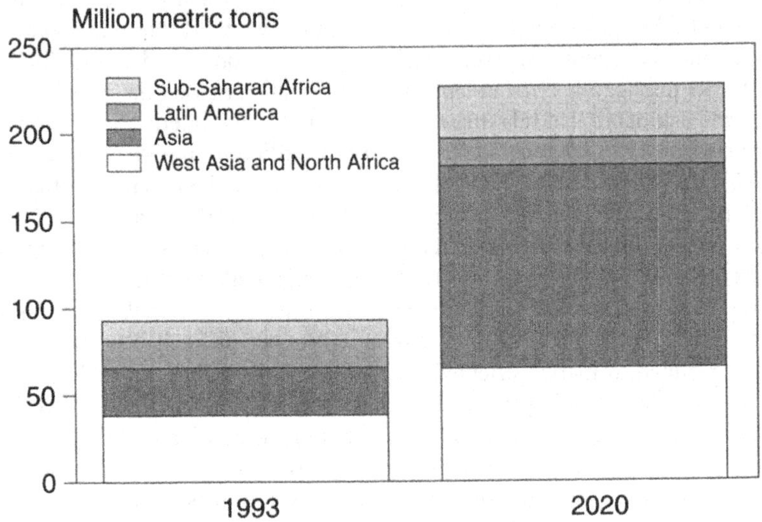

Source: IFPRI IMPACT simulations

tries with 3 billion people may face water stress. In Central Asia, heavy demand for irrigation water is putting a major strain on the Amu Darya River, causing the Aral Sea to shrink. Can cotton farming remain a viable source of income for farmers in Uzbekistan and Turkmenistan over the next two decades without killing off the Aral Sea? Worldwide, our projections call for a 35 percent increase in global water withdrawals over 1995-2020. In developed countries, industry will account for most of the increased demand. In developing countries, household and industrial demand will double at agriculture's expense (Figure 3.4).

One of IFPRI's priorities in this research area focuses on property rights and collective action. These issues affect the way billions of farmers, fishers, and herders manage their lands, water, forests, and other natural resources. Property rights matter because farmers make long-term investments to conserve their natural resources only if they hold secure, long-term rights to their land. Without these rights they cannot be sure they will benefit from any improvements to their land. Households without secure property rights are much more vulnerable to hunger and malnutrition. Collective action is critical because farmers often come together to manage natural resources. Collective action at the local level is essential for some natural resource practices, such as environmentally friendly pest control. IFPRI is leading the efforts of the CGIAR system to study the property rights-collective action-natural resource management nexus. This research may prove helpful to Central Asian policymakers during the transition from mostly state and collective farming to expanded individual land tenure.

FIGURE 3.4 Water withdrawals for domestic, industrial and agricultural uses, 1995 and 2020

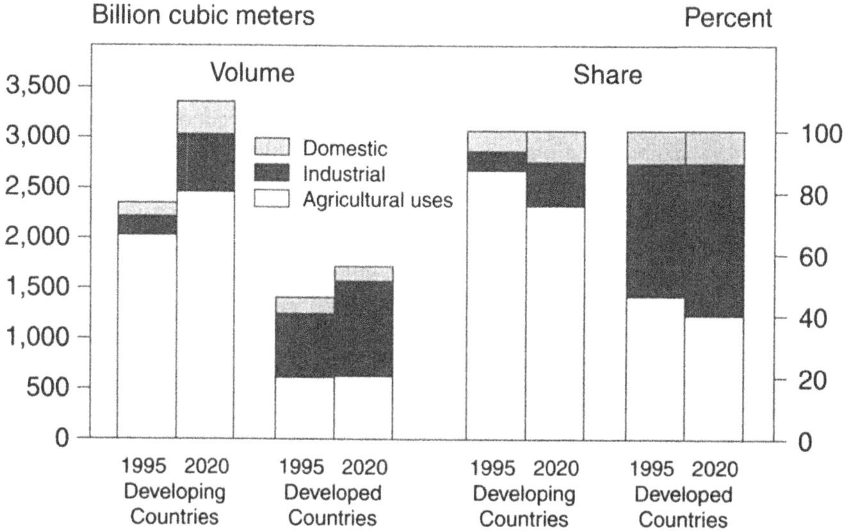

Source: Rosegrant, Ringler, and Gerpacio (1997)

A recent IFPRI study in West Asia and North Africa showed that government policies, together with increasing population and market pressures, have weakened local institutions for managing rangelands as common property. As a result, individual claims to land have increased, farmers have begun to use unsustainable cropping practices on traditional grazing areas, herders have overstocked the area with livestock, and remaining grazing areas are suffering increased degradation. Rehabilitating traditional grazing areas and managing them better in the future will require reforming both agricultural and land policies and transferring ownership and management responsibilities to local communities. But simply reinstating traditional methods that worked well in the past is unlikely to succeed under today's conditions. New approaches to community management need to be tried.

Whereas agricultural development strategies generally emphasize irrigated agriculture and high-potential rainfed areas, a large share of poor, food-insecure people live in the less-favored areas of the developing world. IFPRI research looks at the consequences for productivity, poverty, the environment, and food security of targeting agricultural investment, including agricultural research, to less- and more-favored areas, as well as the appropriate development strategies for different types of less-favored lands. Research has focused on the low-potential rainfed areas of India that are home to tens of millions of poor people, and on semi-arid lands in West Asia and North Africa.

Macroeconomic Policy and Trade Reform

For the past three years, IFPRI's Macroeconomic Reforms and Regional Integration in Southern Africa Project has studied the links between agriculture and the rest of the economy in this subregion. In particular, the project has assessed the effects of macroeconomic policy reforms on income and equity, given the importance of agriculture in terms of its contribution to total economic activity and employment, and given the overwhelmingly rural character of poverty in these countries. This work has relied on two important research tools: social accounting matrices, which provide detailed pictures of the socioeconomic structure of the study countries; and, derived from the latter, computable general equilibrium models for assessing quantitatively the impacts of alternative policies. These research tools might prove valuable for policymakers and analysts in Central Asia.

In the area of global trading arrangements, we have recently produced a set of nontechnical briefs examining the issues facing developing countries in the next round of World Trade Organization negotiations. These should prove useful for a wide range of policymakers, analysts, and interested members of the general public in all global regions. Our research on the impact of further global trade reforms is ongoing.

Food and Nutrition Security

IFPRI's path-breaking research on gender and the intrahousehold aspects of food policy has led to new understanding of the key roles women play as producers and procurers of food, and as guarantors of nutritional security at the household level. IFPRI's work in Bangladesh, in particular, has shown that gender discrimination within households is a critical reason for malnutrition among female children and adults. This research has also shown that well-designed crop diversification programs that are gender-sensitive can help improve incomes and nutrition among poor women and their families.

Despite the rural center of gravity of global poverty, urbanization is proceeding rapidly in the developing world (Figure 3.5), and food insecurity is urbanizing as well. IFPRI research has focused on the appropriate policies for addressing urban food insecurity. These must recognize that city dwellers are much more dependent on income from employment and have fewer opportunities to directly produce their own food. Urban women tend to work outside the home, and questions of access to child care, food safety, and safe sanitation become much more salient. With urbanization, diets often shift from roots, tubers, sorghum, millet, and maize to rice, wheat, and more meat, milk, fruit, vegetables, and processed foods.

IFPRI is leading work by several CGIAR centers on breeding micronutrients such as iron, vitamin A, and zinc into staple crops such as rice, wheat, maize, and beans. Micronutrient deficiencies affect more than 2 billion people, lead to serious

FIGURE 3.5 Urban and rural population in developing countries, 1950–2020

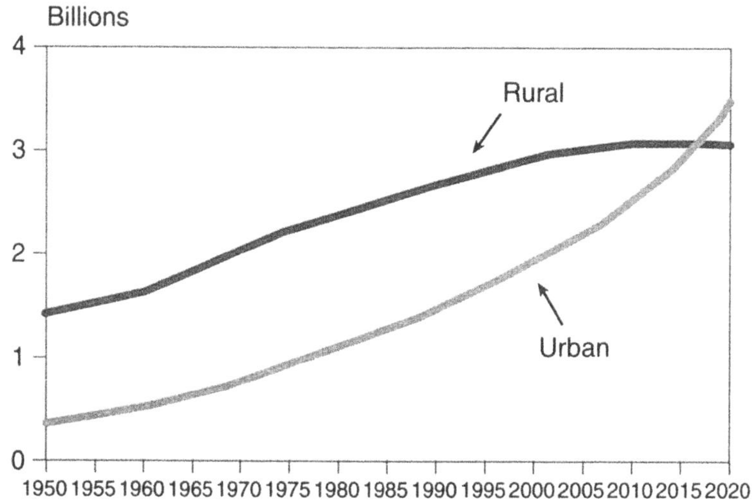

Source: UN (1997)

health problems, and are a significant drain on developing-country economies. Initial results of this research have been very encouraging, and in the long run, breeding is likely to be a more effective route to addressing many types of micronutrient malnutrition than more traditional supplementation or fortification. The work is focusing on crop varieties that are high yielding, as well as micronutrient rich, so that nutrition gains go hand in hand with increased productivity.

The international development community has paid increasing attention in recent years to microcredit as a tool for alleviating poverty. IFPRI research has found that access to credit or participation in credit programs improves household incomes, increases technology adoption, increases food intakes, and provides resources for difficult times. Poor households have high loan repayment rates because they want to be able to obtain additional loans in the future. However, private sector financial institutions remain reluctant to risk offering credit to poor people, so public support in developing appropriate institutions is critical.

As development assistance becomes ever scarcer in the 21st century, it is all the more important that donors and developing country governments make effective use of these resources to promote sustainable poverty alleviation and food security. In Zimbabwe, IFPRI research found that investment in agricultural development provides a more effective buffer against natural disasters than emergency drought relief. Yet too often, emergency aid funds are drawn out of ever-dwindling development assistance resources.

IFPRI research also has focused on improving measures of food security.

Sound policies depend on accurate and comprehensive information. IFPRI is engaged with other international organizations, governments, and nongovernmental organizations in developing the global, regional, national, and subnational food insecurity and vulnerability information and mapping systems that are called for in the World Food Summit Plan of Action. IFPRI, in collaboration with FAO, is seeking to better integrate the results of nationally representative household surveys into the standard measure of chronic undernutrition. Finally, IFPRI is collaborating with the Subcommittee on Nutrition of the United Nations Administrative Committee on Coordination to prepare the fourth *World Nutrition Report*, an authoritative volume on global nutrition issues.

Agricultural Technology

Our work analyzing the policy issues related to funding, managing, and organizing public agricultural research and development (R&D) is well under way. General information analyzed for 22 developed countries shows that public investments in agricultural research slowed substantially during the 1980s compared with the 1970s. Surprisingly, in 1991, developing countries as a group outspent developed countries on public agriculture R&D by $1.1 billion. In developed countries, private spending is growing much faster than public spending. These trends reflect significant changes in the public policy stance toward agricultural R&D, substantial shifts in the private roles in agricultural research, and a changing global context for agricultural research.

Advances in biotechnology—which have been greatly reinforced by evolving intellectual property rights legislation regarding agricultural research—have increased the role of the private sector in research activities which used to be primarily the domain of the public sector. Research outputs are being privatized with stronger intellectual property rights protection; and exchanges of research tools, processes, and relevant technologies held by others are critical for agricultural researchers. Increasingly, the acquisition of proprietary technologies from the private sector is an important consideration for public institutions. Privatization of plant genetic resources poses major policy questions for public agricultural research institutions. Unless CGIAR centers and other public institutions adequately cope with the rapidly emerging new environment, the ability of international public institutions to benefit poor people could be greatly diminished. IFPRI research in this area is focusing on how changing intellectual property rights regimes affect the use and transfer of agricultural genetic resources.

A Shared Vision for the Future

In 1993, in collaboration with partners around the world, IFPRI launched an initiative called A 2020 Vision for Food, Agriculture, and the Environment. It seeks to develop and promote a shared vision and consensus for action for meeting food needs while reducing poverty and protecting the environment. In addition, the initiative seeks to generate information and encourage debate to influence action by national governments, nongovernmental organizations, the private sector, international development institutions, and other elements of civil society. Since 1998, the initiative has increasingly focused on strengthening the capacity of developing countries to conduct their own 2020 research and to design and implement their own 2020 strategies and action plans, mainly through regional networks, initially in East and West Africa, and later in other developing regions. The initiative has distributed more than 400,000 publications worldwide in seven languages, and has established an extremely popular web-site. A 1997 survey of 2020 readers indicated that nearly three-quarters of those surveyed find the publications to be very useful, particularly for formulating policy, preparing policy advice, conducting research, and writing articles or books. The publications and meetings of the 2020 Vision initiative have generated worldwide media attention. 2020 research on China's food needs in the 21st century suggested that China is unlikely to become a major importer or exporter of grain, which helped to defuse alarmist projections. IFPRI's 2020 global projections model is recognized as one of the world's premier models for long-term projections of global food demand, supply, and trade as well as child malnutrition.

Information Sharing and Capacity Strengthening

We are increasingly publishing our research results in formats that are accessible and useful for busy policymakers and the general public. While we continue to publish full-length research reports oriented toward the social science research community, and to publish in scholarly journals, we are also producing policy briefs and longer nontechnical publications that have attracted wide interest among policymakers in both the developing and developed worlds. For example, our research demonstrating that aid to agriculture is a win-win proposition—creating export opportunities for donor countries as well as assisting economic development in recipient countries—has helped encourage some donors to increase their assistance to developing country agriculture.

We seek to integrate capacity strengthening of partners into our research activities. In Mozambique, for example, we worked with the Ministry of Planning and Finance, the Department of National Statistics, and Eduardo Mondlane University on identifying the country's poverty population, assessing the effectiveness of social programs in reaching poor people and reducing poverty, and improving the mechanisms for monitoring poverty. IFPRI undertook training of Mozambican collaborators both in Washington, D.C., and through learning-by-doing involvement in the research. Seminars and conferences were held in Mozambique regarding key research results. IFPRI is developing an operational manual in the local language on the construction of a poverty profile for the country; this will serve as an ongoing training tool for Mozambican policy analysts.

We can achieve the 2020 vision of food security for all, but only if national governments and the international community make a concerted effort to implement appropriate policies. It is essential that food, agriculture, and the environment gain a higher place on the agendas of the world's governments and development institutions. Our dialogue at this meeting is a step toward making the vision a reality. We at IFPRI look forward to fruitful future collaboration with our colleagues in Central Asia.

References

FAO (Food and Agriculture Organization of the United Nations). 1996. Food, agriculture, and food security: Developments since the World Food Conference and prospects. World Food Summit Technical Background Document 1. Rome: FAO.

_____. 1998. The state of food and agriculture 1998. Rome: FAO.

Rosegrant, M. W., C. Ringler, and R. V. Gerpacio. 1997. Water and land resources and global food supply. Paper prepared for the 23rd International Conference of Agricultural Economists, Sacramento, California, 10-16 August.

UN (United Nations). 1997. Urban and rural areas 1996. New York: UN, Department of Economics and Social Affairs.

WHO (World Health Organization). 1998. Nutrition for health and development. <http://www.who.int/nut> (accessed September 17, 1998).

4 Emerging Issues and Challenges for Attaining Food Security in Central Asia

SURESH BABU AND ALISHER TASHMATOV

Since their break up from the Soviet Union, the countries of Central Asia have been undergoing a series of transitions from central planning to market economy. Policy reforms formed a major method of intervention in these countries for the past decade. The process of policy and institutional reforms has been complex and slow; the consensus is that this is largely due to inadequate information for policy planning and formulation (Spoor 1997). In addition, the needed institutional and human capacity to design and implement policy reforms in these countries is severely lacking. The economic crisis precipitated by the disintegration of the former Soviet Union continues after almost a decade of attempts to reform the economies in this region. The viewpoint presented here is based on several rounds of consultations with planners, policy analysts, and program managers in this region. It attempts to synthesize the emerging issues affecting the food and agricultural sectors that have a direct impact on food security, poverty alleviation, and sustainable resource management.

Background

Central Asia consists of five countries—Kazakhstan, Kyrgyz Republic, Tajikistan, Turkmenistan, and Uzbekistan—covering an area of 400 million hectares. This is roughly equivalent in area to Western Europe without Sweden. The region has a population of about 60 million people, growing at an average rate of about 3 percent per year. About 60 percent of the population lives in rural areas, and the numbers of rural workers migrating to urban areas looking for jobs are continuously rising. Agriculture contributes about 30 percent of the region's economy. Wheat and cotton are the major agricultural commodities. The region has a vast area of rangelands (260 million hectares) and livestock production. Meat and wool, in particular, play a major role in food security and income generation. A climate with low rainfall and extreme temperatures, and a landscape of mountains, deserts, and steppes characterize the region. Geographically, Central Asia represents a strategic crossing point, and its economic development will have implications for Russia and the other countries of the former Soviet Union as well as Eastern Europe, China, the Middle East, and South Asia.

The Central Asian republics are currently facing serious development challenges associated with increasing food security, alleviating poverty, and minimizing natural resource degradation. The past decade's transition from a centrally planned economy toward a market-oriented economy has not been adequately supported by institutional development. As a result, the standard of living has declined. Although detailed information on the status of poverty and food insecurity is not available, anecdotal evidence suggests that food insecurity caused by declining incomes in recent years is a major development concern in these countries. Information on the incidence of malnutrition is scanty. However, available data indicate that child malnutrition in these countries is widespread. For example, in a rural district in Uzbekistan, about 40 percent of children under age three are stunted, while one in five children is stunted in Tashkent, the capital city. Similarly, in the Varzob district in Tajikistan, it has been reported that about 54 percent of children are stunted (Peimani 1997). Although not much is known about the condition of the natural resource base, available information suggests that poor design and slow implementation of institutional and policy reforms have resulted in degradation of natural resources, threatening the sustainability of agricultural and livestock production.

Reversing these trends in Central Asia will require sustained policy action over a long period. Given the lack of adequate information on the magnitude of and trends in food insecurity, poverty, and natural resource degradation, there is an urgent need to assess these conditions for policy planning and formulation. In addition, supporting the exchange of information and the conduct of policy dialogue among policy researchers, decisionmakers, and members of the civil society in this region is essential to identify and prioritize policy information and research needs and place these issues on the policy agenda of these countries. Developing long-term strategies for improving food security, alleviating poverty, and encouraging the sound use of natural resources is fundamental to providing an appropriate policy direction for sustained economic growth in the region. To achieve these objectives, several emerging issues and challenges must be adequately recognized.

Emerging Issues and Challenges

Several emerging issues warrant policy attention in this region. Since the break up of the Soviet Union, this region has faced special challenges. This is particularly evident in the rural sector, where the appropriate interface of technological change, institutional development, and policy reform is largely missing. In Central Asian countries, food and agricultural sectors represent both the promise and the challenge for the future. The role agriculture will play in these developing economies is unclear. The extent to which agriculture will remain a major supplier of raw materials and an important source of export revenue is also unclear. However, the experience of other developing countries shows that sustained growth in the agricultural sector is fundamental for stimulating income growth and demand for industrial and manufactured products. Transforming the agricultural sector into a market-based system in the face of continuing economic instability remains a major challenge in Central Asia. For example, before the break up of the Soviet bloc, the central planning system supplied the region with agricultural inputs—such as seeds, fertilizers, and pesticides—from outside the region. With the help of these modern inputs, the region specialized in the production of commodities such as cotton and wheat. The supply of inputs stopped abruptly after the break up. As a result, agricultural production continues to suffer from low levels of commercial inputs. In spite of these constraints, for the majority of the rural population, agriculture and livestock production remain the strongest basis for food security and poverty alleviation in Central Asia. Thus one of the major challenges to the policymakers in this region is to identify the nature and the direction of policy incentives that will put the agricultural sector on a dynamic growth path.

Reorientation of crop and livestock production from collective state farms to individual farms without adequate institutional arrangements has destabilized agricultural production systems. Land reforms have not been completed and continue to pose challenges for increasing yield levels which are well below those of other members of the Commonwealth of Independent States (Beniwal, Suleimenov, and Srivastava 1998). Further, land reforms have not been adequately accompanied by other market reforms. This has resulted in poor management of land and water resources. Developing appropriate institutions to support the process of land and market reforms and identifying the environment that will enable these institutions to evolve in response to policy changes remain major development challenges.

Irrigation systems that were developed for collective state farms are not efficient for individual farm management. Although water use associations are being formed to regulate and price water use, irrigation systems continue to operate inefficiently because of individual decsionmaking and the crop choices of farmers. For example, nearly all of the arable land (1.4 million hectares) in Turkmenistan is irrigated and faces environmental problems of salinity and water logging because of inappropriate and excessive irrigation (O'Hara 1997). Policy incentives that will ensure efficient use of water resources need to be identified. Lessons and experiences from south and west Asian countries which have similar institutional and resource challenges would be helpful in identifying appropriate strategies in this regard.

Monetization of new republics with low levels of foreign exchange, coupled with poor trading arrangements (which slow down transactions and payments for the products exported from the region), has created economic uncertainties. For example, Turkmenistan continues to export natural gas to other members of the Commonwealth of Independent States although it has not received full payments for the past several years, including some years during the former Soviet era. Low levels of foreign exchange for importing food grains and other food commodities have forced these countries to focus on foodgrain self-sufficiency as a food security strategy although that may not be economically efficient. The food self-sufficiency objective has major implications for resource allocation in terms of land and water use in the region. Thus identifying the appropriate trading arrangements among the countries in the region and with countries outside the region is essential for enhancing the efficient use of resources.

In addition, uncertainty in regional trade for food commodities and the resulting focus on food self-sufficiency as a strategy for food security have also forced farmers to concentrate on domestic production of cereals. This has resulted in monoculture of wheat production without adequate use of enhanced fertilizers. New areas have been brought into cereal cultivation. For example, the area under wheat cultivation has increased 25 percent since 1990. These factors have contributed to a significant reduction in the levels of soil fertility. In addition, monoculture of cotton, the major cash crop, has resulted in soil nutrient mining in the major cotton growing areas of the region because of the lack of crop rotation. The reduction in the trade flow for agricultural inputs between the Central Asian republics and the rest of the world has also diminished the availability of feedgrain for livestock production, which has increased the pressure on rangelands (Wilson 1997).

Recently, Central Asian countries have made varying degrees of progress toward membership in the World Trade Organization (WTO). How their accession to WTO will affect the region's trade, environmental, and food security linkages remains unclear. The migration of the rural unemployed to urban areas is rising in all the Central Asian countries, increasing the food insecurity and poverty among the urban population. Government policies have failed to reverse this trend, so the governments have been forced to identify social protection policies to provide safety nets to the poor and vulnerable in the urban areas. Thus there is an urgent need to identify policies that ensure food security for the rural and urban poor and to assist the governmental and nongovernmental agencies in designing and implementing innovative programs targeted at poverty alleviation.

At the regional level, the continual drying of the Aral Sea and the resulting creation of vast salt beds pose a serious environmental threat to the water resources of the region. It is feared that salty winds from the dried up parts of the Aral Sea will contaminate the glaciers of the Central Asian mountains and render the water resources saline. Increasing deterioration of the mountain farming systems through soil erosion and soil degradation in rangelands resulting from overgrazing and deforestation are issues common to all the countries in the region. Thus there is a need to develop a common understanding and consensus on the regional resource problems that affect the region as a whole (Kharin 1996).

Informing the policymakers on the potential impact of alternative policies is central to the policy reform process. Designing and implementing successful policy reforms requires sound information on the various sectors of the economy. Adequate institutional development and appropriate capacity for information generation are critical. While policy changes need to be initiated and formulated by the individual countries, the effect of policies and programs implemented for the past decade should be understood. In addition, monitoring the changes in the welfare of the population will require an adequate database on the indicators and causal factors associated with food security, poverty and sustainable use of natural resources.

What Needs to Be Done?

These are only a few of the issues facing the food, agricultural, and natural resource sectors, and an inventory of the major challenges facing policymakers in the region is urgently needed. Identifying the information gaps for the formulation and implementation of policy reforms is essential. The region also needs information on the experiences of other developing countries as it faces the newly emerging challenges in the transition to market-oriented economies.

The region must also develop an understanding of and consensus on the problems it faces and the actions needed to alleviate poverty, reduce food insecurity, and substainably manage the natural resource base in Central Asia. This could be achieved by bringing together policymakers and decisionmakers in government ministries such as agriculture, rural development, environment, natural resources, and finance and planning, along with policy analysts, advisers, and influential members of civil society such as nongovernmental organizations and researchers from each of the countries in the region. Thus, a forum for sharing experiences among the countries with similar institutional and policy constraints would be of immediate help. Poverty, food insecurity, and natural resource degradation are closely intertwined in a self-perpetuating negative spiral, and these issues cut across all the countries of Central Asia. Each country is attempting to arrive at solutions to these problems in its own way, but there are considerable synergies to be gained by sharing strategies and experiences from countries in other parts of the world that have similar challenges.

The issues and challenges identified in this paper are not totally new. They have been recognized by various authors who have studied the evolution of these economies for the past decade (Duncan 1994). However, development of a regional long-term strategy for poverty alleviation and sustainable use of natural resources remains elusive. Institutions must be developed that will foster a policy dialogue on long-term issues related to food, agriculture, and the environment within and among these countries. Such institutions will be fundamental to identifying development strategies for the next 10 to 15 years. Priorities must be set for future food and agricultural policy research and analysis that will help in generating and sharing information useful to Central Asia in its quest to eradicate poverty and protect its natural resource base.

References

Beniwal, S. P. S., M. Suleimenov, and J. P. Srivastava. 1998. Trends in agricultural production and research in the transition economy of the Kyrgyz Republic. Ankara, Turkey: International Center for Agricultural Research in Dry Areas.

Duncan, A. 1994. Agricultural and economic reform issues in Kyrgyzstan, former Soviet Central Asia, Food Policy 19 (1): 85-87.

Kharin, N. 1996. Strategy to combat desertification in Central Asia. Desertification Control Bulletin 29: 24-29.

O'Hara, S. L. 1997. Irrigation and land degradation: Implications for agriculture in Turkmenistan, Central Asia. Journal of Arid Environments 37 (1): 165-179.

Peimani, H. 1997. Stunting in Central Asia and the Caucasus. In World Nutrition Situation. Geneva: Administrative Committee on Coordination/ Subcommittee on Nutrition.

Spoor, M. 1997. Agrarian transition in the former Soviet Union: The case of Central Asia. In The Market Panacea: Agrarian transformation in developing countries and former socialist economies, ed. M. Spoor. London: Intermediate Technology Publications.

Wilson, R. T. 1997. Livestock, pastures, and the environment in Kyrgyz Republic, Central Asia. Mountain Research and Development 17 (1): 57-68.

5 Food Policy Research for Improving the Reform of Agricultural Input and Output Markets in Central Asia

FRANCESCO GOLETTI AND PHILIPPE CHABOT

Agriculture is a significant component of the economies of Central Asia. Before 1992, the Soviet government invested substantially in the sector, through such programs as the Virgin Lands campaign in Northern Kazakhstan in the 1950s and 1960s, and the spectacular expansion of irrigation in the Aral Sea Basin. The sector currently accounts for between 10 and 45 percent of gross domestic product (GDP) among the republics and employs 22 to 49 percent of the labor force (World Bank 1998b). Because of this, and because of its export potential, the sector will continue to be an important source of growth in the foreseeable future. Since independence, however, production has fallen sharply owing to shocks stemming from the Soviet breakup. The republics have implemented a range of policies to address the transition to a market-based economy and to foster agricultural growth. Central to this transition is the reform of agricultural input and output markets. The objective of this chapter is to discuss some important policy issues that research should address to improve the reform of these markets. The chapter is organized into three main sections. First, we describe some general features of the agricultural sector in each of the region's economies and briefly discuss trends since independence. Second, we address some policy issues affecting the reform of input and output markets. Third, we outline areas which warrant further research.

The Agricultural Sector in Central Asia

Despite key differences in the economies of Central Asia, they share a common reliance on agriculture as a key source of employment and foreign exchange. Macroeconomic performance across the region has been very poor since independence. The agricultural sector has contracted too, but to a lesser extent than aggregate output. It has also undergone a reorientation of its cropping pattern from cash crop to cereal production. In the sections that follow we will provide evidence on the importance of agriculture for the region, summarize the macroeconomic and agricultural performance since 1992, and give some details about the shifting pattern from cotton to wheat production.

The Importance of Agriculture

Agriculture is a crucial source of regional employment; except in Kazakhstan, it occupies nearly half of the republics' labor force (Table 5.1). The share of agriculture in GDP ranges from 10 to 45 percent (Table 5.2). Since 1992, the sector has grown as a component of GDP in the Kyrgyz Republic and Tajikistan, while it declined slightly in Turkmenistan and Uzbekistan, and substantially (by 55 percent) in Kazakhstan.

TABLE 5.1 Central Asia, basic indicators in 1997

	Per capita GDP	Percentage of labor force in agriculture
Kazakhstan	1,340	22
Kyrgyz Republic	440	48
Tajikistan	330	49
Turkmenistan	630	44
Uzbekistan	1,010	42
Averages		
Low-income countries	350	N/A
Lower middle-income countries	1,230	N/A

SOURCE: World Bank 1999.

NOTE: N/A means not applicable.

TABLE 5.2 Agriculture as a percentage of GDP, 1992-97

	1992	1993	1994	1995	1996	1997	1997 as a ratio of 1992
Kazakhstan	27	18	15	13	13	12	.44
Kyrgyz Republic	39	41	41	44	50	45	1.15
Tajikistan	n.a.	21	19	15.3	27.7	27.6	1.31
Turkmenistan	n.a.	12	9	6	6	10	.83
Uzbekistan	41	32	39	32	26	31	.75

SOURCE: World Bank 1999.

NOTE: n.a. indicates not available.

Moreover agriculture is a crucial and growing source of foreign exchange (Table 5.3). The share of agriculture in total exports ranged from 12 percent in Kazakhstan to 39 percent in Tajikistan in 1997. Its share in exports grew in each republic from 1993 to 1997. Growth varied from 6 percent in Kazakhstan to 26 percent in the Kyrgyz Republic.

In the early 1990s, agricultural imports exceeded exports in Kazakhstan, the Kyrgyz Republic, and Tajikistan, resulting in a trade deficit for the agricultural sector. However, strong increases in agricultural exports moved the trade balance into surplus in each republic by 1996 (Table 5.4). Agricultural imports have generally remained constant or fallen while agricultural exports have increased significantly in all the republics except Turkmenistan. This sectoral trade surplus has served to bolster the overall balance of payments situation in each country. Thus agricultural trade has emerged as a key earner of foreign exchange and its potential for further growth is high.

Macroeconomic and Agricultural Performance since 1992

Gross domestic product. All the Central Asian economies contracted significantly following the Soviet breakup. Economic growth was negative for the republics until 1995 (Table 5.5). However, GDP increased in Kazakhstan, the Kyrgyz Republic, and Uzbekistan in 1996, and more strongly still in 1997. In Tajiskistan and Turkemenistan, GDP continued to fall in 1996. In 1997, no data are available for Tajikistan, but owing to an oil shock, GDP in Turkmenistan plummeted by a remarkable 26 percent.

TABLE 5.3 Agricultural exports as a percentage of total exports, 1993-97

	1993	1994	1995	1996	1997	Change 1993-97
Kazakhstan	6	6	9	11	12	6
Kyrgyz Republic	5	10	14	38	31	26
Tajikistan	4	8	11	n.a.	n.a.	3a
Turkmenistan	15	19	23	22	39	24
Uzbekistan	4	18	23	27	26	22

SOURCES: Calculated from FAO various years; World Bank 1999; and IMF 1998a, 1998b, 1998c, 1998d, and 1999.

NOTE: n.a. indicates not available.

[a] This number is the change between 1994 and 1995.

TABLE 5.4 Trade in agricultural goods, 1992-97 (millions of U.S. dollars)

	1992	1993	1994	1995	1996	1997	1997 as a ratio of 1992
Agricultural imports							
Kazakhstan	612	432	501	498	487	603	.99
Kyrgyz Republic	290	204	91	105	181	112	.38
Tajikistan	282	319	213	221	146	183	.64
Turkmenistan	282	267	158	196	262	277	.98
Uzbekistan	974	692	827	781	897	804	.83
Agricultural exports							
Kazakhstan	385	630	471	661	810	922	2.39
Kyrgyz Republic	59	67	104	137	213	207	3.51
Tajikistan	77	76	126	207	175	196	2.55
Turkmenistan	455	393	404	481	368	369	.81
Uzbekistan	1,264	672	1,719	1,992	1,987	1,767	1.40
Net agricultural exports							
Kazakhstan	- 227	198	- 30	162	323	319	
Kyrgyz Republic	- 231	- 137	13	32	32	95	
Tajikistan	- 214	- 242	- 87	- 14	29	14	
Turkmenistan	173	125	246	285	106	92	
Uzbekistan	290	- 19	891	1,211	1,090	962	

SOURCE: FAO various years.

TABLE 5.5 Percentage change in gross domestic product, 1992-97

	1992	1993	1994	1995	1996	1997	Yearly average 1992-97
Kazakhstan	-13	-10	-13	-8	0	2	-7.0
Kyrgyz Rep.	-16	-16	-20	-5	7	10	-6.7
Tajikistan	-30	-11	-22	-12	-4	n.a.	-15.8
Turkmenistan	-5	-10	-19	-9	-8	-26	-15.4
Uzbekistan	-11	-2	-4	-1	2	5	-1.8

SOURCES: World Bank 1999; and IMF 1998a, 1998b, 1998b, 1998d, and 1999.

NOTE: n.a. indicates not available.

Output. Table 5.6 provides an index of macroeconomic output for all five Central Asian republics compared with the average of the former Soviet Union (FSU). Kazakhstan, the Kyrgyz Republic, and Uzbekistan have experienced better rates of aggregate output than the FSU average. Turkmenistan outperformed the FSU average until 1996, when a shock to its gas sector had a severe negative impact. Since 1991, Tajikistan has performed below the FSU average, owing to civil strife.

Of the Central Asian republics, Uzbekistan has experienced the smallest contraction in total output. Indeed its contraction is the smallest of any of the former Soviet epublics. Observers cite two major reasons for this. The first is Uzbekistan's relatively low level of industrialization before 1992, and the commensurately low degree of integration of its industrial sector with other parts of the FSU. The second is that Uzbekistan was more able than the other former Soviet republics to find output markets outside of the Commonwealth of Independent States (CIS) for its traditional exports of natural gas and cotton (Taube and Zettlemeyer 1998).

Agricultural production. In line with aggregate trends, the agricultural sector also contracted in each of the republics following independence, but less so than total GDP.

Except in the Kyrgyz Republic, the production of cotton, the primary export crop, contracted significantly between 1992 and 1998 (Table 5.7). This is partly the result of shocks stemming from the economic transition. One clear manifestation is dramatic declines in the use of such inputs as fertilizer and farm machinery (Tables 5.8 and 5.9).

In contrast to cotton, grain production has grown among the Central Asian republics over the past six years, except in Kazakhstan (Table 5.10). In the rest of the region, it grew between 190 and 320 percent.

In Kazakhstan, by contrast, grain production declined dramatically. In 1998, it was only 26 percent of its 1992 level, which was already below pre-independence levels. Much of the land that had been cultivated to grain earlier was not

TABLE 5.6 Index of total output, 1991-97 (1991 = 100)

	1991	1992	1993	1994	1995	1996	1997
Kazakhstan	100	95	85	74	68	68	70
Kyrgyz Republic	100	86	73	58	55	58	62
Tajikistan	100	71	63	50	43	31	32
Turkmenistan	100	95	85	69	64	62	47
Uzbekistan	100	89	87	83	82	84	86
FSU average	100	81	72	62	59	59	60

SOURCE: Taube and Zettlemeyer 1998.
Note: FSU indicates the former Soviet Union.

appropriate for this activity, explaining part of the decline. However, this factor alone cannot explain the tremendous decline in area, nor the sharp fall in yields (Table 5.11). As with cotton, economic disruptions wrought by the transition are a crucial explanatory factor.

The contraction of the grains sector has severely affected rural producers in Kazakhstan. Moreover, observers disagree as to Kazakstan's future potential.

TABLE 5.7 Seed cotton production, 1992-98 (thousands of metric tons)

	1992	1994	1996	1998	1992-98
Kazakhstan	246	208	183	162	.66
Kyrgyz Republic	52	54	73	75	1.44
Tajikistan	515	531	318	385	.75
Turkmenistan	1,290	1,283	436	707	.55
Uzbekistan	4,128	3,936	3,350	3,220	.78

SOURCE: FAO various years.

TABLE 5.8 Fertilizer application rates, 1992-96 (kilograms per hectare)

	1992	1993	1994	1995	1996	1996/1992 ratio
Kazakhstan	13	9	3	3	4	.30
Kyrgyz Republic	26	21	21	22	23	.89
Tajikistan	149	87	82	88	91	.61
Turkmenistan	130	110	78	81	87	.66
Uzbekistan	163	142	106	106	118	.73

SOURCE: FAO various years.

TABLE 5.9 Tractors per 1,000 hectares, 1992-97

	1992	1993	1994	1995	1996	1997	1997/1992
Kazakhstan	6.2	6.0	5.6	5.3	4.6	3.6	.58
Kyrgyz Republic	20.0	19.3	20.2	19.8	14.5	14.1	.71
Tajikistan	41.5	34.5	35.5	37.5	39.5	39.5	.95
Turkmenistan	46.5	36.1	31.3	30.8	30.7	30.7	.66
Uzbekistan	40.2	38.0	38.0	38.0	38.0	38.0	.95

SOURCE: FAO various years.

Some feel that the country holds great promise in the medium run as a major export earner and source of economic growth. Others argue that because of risky growing conditions and high transportation costs for grain to reach major markets, the country holds little future as a major grains producer. Further work is needed to establish the relative value to be derived from grain production versus other rural activity. With the exception of Turkmenistan, the Central Asian countries have not experienced such sharp decreases in wheat productivity. Indeed, for the most part, wheat yields have either remained constant or increased over the past seven years.

Cotton yields have declined throughout the region over the past decade (Table 5.12). This is in part a result of disruptions created by economic transition. However, ecological or environmental factors may also be responsible.

Livestock. The livestock sectors in Kazakhstan, the Kyrgyz Republic, and Tajikistan have contracted sharply since independence (Tables 5.13 and 5.14). In Kazakhstan, the change is most severe: in 1998, herd numbers for cattle and sheep were at only 44 percent and 26 percent of their respective 1992 levels.

TABLE 5.10 Wheat production, 1992-98 (thousands of metric tons)

	1992	1994	1996	1998	1998/1992 ratio
Kazakhstan	18,285	9,052	7,678	4,746	.26
Kyrgyz Republic	679	608	1,040	1,290	1.90
Tajikistan	156	182	395	470	3.01
Turkmenistan	377	675	424	600	1.59
Uzbekistan	964	1,362	2,742	3,094	3.21

SOURCE: FAO various years.

TABLE 5.11 Wheat yields, 1992-98 (tons per hectare)

	1992	1994	1996	1998	1998/1992 ratio
Kazakhstan	1.32	.72	.63	.52	.40
Kyrgyz Republic	2.73	1.83	2.30	2.77	1.01
Tajikistan	.85	1.06	1.43	2.03	2.38
Turkmenistan	1.91	2.58	.73	1.21	.63
Uzbekistan	1.54	1.42	2.06	3.21	2.09

SOURCE: FAO various years.

Uzbekistan and Turkmenistan have been considerably more successful in maintaining livestock numbers but this may be a function of continued state support.

The large declines in Kazakhstan, the Kyrgyz Republic, and Tajikistan can partly be explained by problems associated with economic liberalization. However, for Kazakhstan, and also in the Kyrgyz Republic, the decrease in herd numbers was also due to problems of financial liquidity in the sector. After inde-

TABLE 5.12 Cotton yields, 1992-98 (tons per hectare)

	1992	1994	1996	1998	1998/1992 ratio
Kazakhstan	2.20	1.88	1.72	1.41	.64
Kyrgyz Republic	2.44	2.02	2.31	2.34	.96
Tajikistan	1.81	1.88	1.39	1.57	.87
Turkmenistan	2.28	2.30	.84	1.22	.54
Uzbekistan	2.48	2.56	2.25	2.10	.85

SOURCE: FAO various years.

TABLE 5.13 Cattle production, 1992-98 (thousands of head)

	1992	1994	1996	1998	1998/1992 ratio
Kazakhstan	9,084	9,347	6,860	3,999	.44
Kyrgyz Republic	1,190	1,062	869	830	.70
Tajikistan	1,221	1,250	1,147	1,040	.85
Turkmenistan	777	1,104	1,199	900	1.16
Uzbekistan	5,113	5,431	5,204	5,300	1.04

SOURCE: FAO various years.

TABLE 5.14 Trends in sheep production, 1992-98 (thousands of head)

	1992	1994	1996	1998	1998/1992 ratio
Kazakhstan	33,908	33,312	18,786	8,908	.26
Kyrgyz Republic	9,225	7,103	4,075	3,350	.36
Tajikistan	2,584	2,078	1,805	1,600	.62
Turkmenistan	5,380	6,000	6,150	5,400	1.00
Uzbekistan	8,274	9,360	8,352	8,000	.97

SOURCE: FAO various years.

pendence, the sale of livestock was often the only means for producers to obtain cash to pay wages and procure farm inputs. This liquidity constraint was so severe that for a protracted period, producers continued to market their livestock with little regard to output price. This resulted in a glut of livestock on the market, lowering prices, and a large decrease in aggregate herd numbers.

Some have argued that the fall in livestock may also reflect a transition toward activities that are more remunerative in the new economic climate. They point out that before independence, agricultural production in the Soviet Union heavily emphasized meat and dairy production, and that per capita meat consumption was quite high relative to other countries at similar stages of development (Table 5.15). Nevertheless, given the significant pasture resources in Kazakhstan, the Kyrgyz Republic, and Tajikistan, livestock appears to hold considerable promise as a source of aggregate growth and of rural household income.

Fertilizer and farm machinery. Farm input use declined dramatically in Central Asia over the past six years. No data is available on fertilizer and tractor use before 1992, but it is generally believed that agricultural productivity in Soviet Central Asia was quite low. Levels of inputs used compared with other countries were generally very high, particularly in relation to output. Again, part of the subsequent decline may be due to an adjustment process to reflect relative scarcity of factors of production.

Shifts in Cropping Patterns

Since 1992, most Central Asian republics have reoriented their production away from cotton and toward wheat. In fact in Uzbekistan and Turkmenistan, the state has directed that land be planted to wheat in place of cotton. This emphasis on self-sufficiency is probably a reaction to severe economic contraction and uncertainty regarding the future. It may also result from a lack of marketing infrastructure to adequately dispose of cotton in a liberalized economy.

Table 5.16 shows the production of wheat as a share of cotton between 1992 and 1998. It reveals that while the proportion of cotton to wheat remained rough-

TABLE 5.15 Per capita meat consumption in the Central Asian republics, Turkey, and Iran, 1992 and 1997 (kilograms per year)

	Kazakhstan	Kyrgyz Republic	Tajikistan	Turkmenistan	Uzbekistan	Turkey	Iran
1992	55.1	48.1	15.0	35.3	24.7	19.4	22.0
1997	45.0	35.8	8.5	26.7	27.4	19.6	23.4

SOURCE: FAO various years.

TABLE 5.16 Cotton production as a share of wheat production, 1992-98 (thousands of metric tons)

	1992	1994	1996	1998	1998/1992 ratio
Kazakhstan	1.35	2.30	2.38	3.41	2.07
Kyrgyz Republic	7.66	8.88	7.02	5.81	-1.84
Tajikistan	330.13	291.76	80.51	81.91	-248.21
Turkmenistan	342.18	190.07	102.83	117.83	-224.34
Uzbekistan	428.22	288.99	122.17	104.07	-324.14

SOURCE: FAO various years.

ly constant in Kazakhstan and the Kyrgyz Republic, it decreased substantially in the other three republics by 224 to 324 percent.

It is not clear that this shift toward cereals is the most efficient use of these countries' agricultural resources. Before independence, the Central Asian republics met their grain needs by importing grain from other parts of the Soviet Union, notably Russia and Kazakhstan. Increased grain production in place of cotton may reduce environmental pressure because it requires less irrigation water and agrochemicals than does cotton. However, given each country's resource endowment, it may be more beneficial for each country to continue or even increase its imports of grain from traditional exporters and use its own land for cash crop production. It is not yet clear whether Central Asian land is best suited to grain or cash crop production. Further research could help determine each republic's areas of comparative advantage.

Issues in Agricultural Input and Output Market Reforms

The preceding discussion provided an overview of developments in the agricultural sector of Central Asia since independence. However, to understand these developments better, it is necessary to review policies the Central Asian republics have adopted to manage the economic transition. The republics have pursued substantially different degrees and types of reform. While certain countries (Kazakhstan, the Kyrgyz Republic, and Tajikistan) have aggressively pursued liberalization policies designed to transform their systems of production, others (Uzbekistan and Turkmenistan) have been more inclined to maintain economic structures similar to those in place before independence. In spite of a movement toward a market-based system, several constraints remain in Central Asian agricultural input and output markets, such as price controls for cotton and fertilizer subsidies. Moreover, the difficulties inherent in removing these constraints are compounded by the difficulty of how to best sequence reforms in agricultural input and output markets and the task of overcoming institutional weaknesses. This section focuses on price policy affecting agricultural input and output markets and institutional constraints in water management, seed multiplication, and research.

Price Controls in Cotton

The degree of liberalization in input and output markets is highly correlated with the degree of liberalization in macroeconomic policy. This relationship is evident in the market for cotton. Although all the Central Asian republics initially relied on state procurement of cotton, three have since undertaken liberalization measures. The Kyrgyz Republic eliminated state procurement in 1992, a step which was entirely in keeping with its larger macroeconomic reform program. Kazakhstan and Tajikistan soon followed. In contrast, both Uzbekistan and Turkmenistan have yet to lift procurement policies for cotton. This is in line with their much slower approach to market-oriented macroeconomic reform.

A comparison of procurement policies for cotton warrants closer scrutiny because it shows the effect such policies can have on producer prices. Table 5.17 shows the output price for seed cotton in 1997 for all Central Asian republics in both local currency units and in U.S. dollars. It shows that output prices for producers in Kazakhstan, the Kyrgyz Republic, and Tajikistan are similar, and are relatively close to the average estimated border parity price for cotton.

Since there is no state procurement price in these countries, the border parity price can be expected to roughly equal the producer price if the cotton market in the country is efficient. Producer prices are also not lowered by an inflated offi-

TABLE 5.17 Output price for cotton, 1997

	Kazakhstan	Kyrgyz Rep.	Tajikistan	Turkmenistan	Uzbekistan	Average estimated border parity price
Output price/ton (LCU)	25,500	7100	190,000	1,000,000	14,750	
Output price/ton (USD)	349	394	388	240	242	404
Output price at parallel exchange rate	349	394	388	188	105	404

SOURCES: IMF 1998a, 1998b, 1998c, 1998d, and 1999; USDA various years; and EU-TACIS 1995.

NOTE: LCU is local currency units and USD is U.S. dollars.

cial exchange rate. In contrast, in both Uzbekistan and Turkmenistan, the state continues to procure cotton at regulated prices. As shown in the Table 5.17, producer prices are far below those obtained in Kazakhstan, the Kyrgyz Republic, or Tajikistan, and below the estimated border parity price. Moreover, in Uzbekistan and Turkmenistan there are significant differences between the official exchange rate and the shadow exchange rate. Owing to this gap, producers receive even lower farmgate prices than if the market determined the exchange rate.

Such a procurement policy acts effectively as a tax on producers equal to the difference between the price paid to the producer and the border price. It is frequently argued that producers are compensated for these transfers by state subsidies on inputs and by the state's provision of free or cheap irrigation water. However, analyses show that the net effect of the transfers on producers is negative and sizeable. In the case of Uzbekistan, this transfer is estimated at between 3 and 7 percent of GDP (IMF 1998d). Moreover, producer output falls below what it would be if they were to receive market prices.

More generally, the historical record shows that economies with large amounts of state intervention in agriculture tend to perform poorly. A significant literature demonstrates that a highly interventionist strategy is not the most effective way to develop a dynamic agricultural sector (Krueger, Schiff, and Valdes 1992). Accordingly, efficient market reform would suggest the removal of state subsidies and price controls within both input and output markets.

Sequencing of Market Reforms in the Wheat Sector

Similar procurement policies and distortions are evident in regional grain markets too. Here we will consider the effects of grain market liberalization in Kazakhstan, a case which illustrates some of the potential problems policymakers should consider in implementing agricultural reforms.

Kazakhstan eliminated its procurement policy for grain in 1995. However, prices for farm inputs such as fuel and fertilizers were liberalized starting in 1993. Owing to this sequencing gap, the grains sector experienced a severe trade shock over a two-year period. This process is shown clearly in Table 5.18, which tracks the price of wheat relative to other farm inputs from 1990 to 1995.

Because producers were forced to pay market prices for inputs while receiving below-market prices for outputs, they incurred significant quantities of debt as they sought to meet state procurement orders for wheat. One result is that many producers in Kazakhstan remain mired in debt. This example demonstrates that while the end effects of liberalization policies can be positive, it is crucial to consider such factors as their sequencing.

Terms of trade improved once output prices for grain were liberalized in 1995 (Table 5.19). Even so, producers continue to contend with the country's geographic remoteness and climatic variation. Lack of infrastructure may add tremendously to production costs. One study found that depending on production and market conditions in both Russia and Europe, the estimated farmgate price of grain in Kazakhstan can range between $40 and $180 per ton (EU-TACIS AgroInform 1997).

Moreover, the grain marketing sector in Kazakhstan is beset with significant impediments resulting from a lack of credit, ineffective contracting mechanisms, and a lack of market information, all of which serve to lower the farmgate producer price. Further research is needed to more clearly illustrate these constraints and to propose ways in which they can be overcome.

Fertilizer Distribution and Subsidies

As already noted, fertilizer application rates declined precipitously during the past decade in Central Asia. Before independence, fertilizer application rates in Soviet Central Asian agriculture were very high, so a decrease in usage may have been positive in mitigating soil nutrient balance. Bolstering the region's productivity, however, will require an adequate fertilizer supply and distribution system. In Turkmenistan and Uzbekistan, where the state continues to control the fertilizer supply, producers have not reported significant difficulties in obtaining it. Producers in Kazakhstan, the Kyrgyz Republic, and Tajikistan, in contrast, frequently report inadequate supplies.

Uzbekistan is the major producer of nitrogenous fertilizer, a by-product of its substantial natural gas industry. It exports approximately 25 percent of its sup-

TABLE 5.18 Terms of trade in Kazakhstan: Tons of wheat required to purchase each item

	1990	1991	1993	1995
Tractor	76	150	367	310
Combine	50	50	230	580
Fertilizer (1 ton)	3	2	22	2
Fuel (1 ton)	1	2	30	3

SOURCE: World Bank 1994.

TABLE 5.19 Output prices for grains in Kazakhstan (U.S. dollars per ton)

	1992	1993	1994	1995	1996	1997	1998
Price at Kazakhstan border	30.5	42	45	75-110	110-120	100-120	90-110

SOURCE: USDA various years.

ply (Tables 5.20 and 5.21). Though Kazhakstan produces just 10 percent of Uzbekistan's total, it exports 50 percent of its supply.

Before 1992, production of phosphate fertilizers was concentrated in an area shared by Southern Kazakhstan and Uzbekistan. Since independence, however, its production has declined sharply in both countries (Table 5.22). It fell to 20 to 40 percent of its respective 1992 levels by 1994, then grew slightly thereafter.

Russia was and continues to be the major supplier of potash fertilizer to the Central Asian republics. Surprisingly, though it is imported, its consumption has not declined very much (Table 5.23). Data suggests that imports of this fertilizer have continued at a relatively stable rate throughout the region and have actually increased in Turkmenistan and Uzbekistan. This could be because potash fertilizer from Russia is bartered in exchange for such products as oil and natural gas.

As with farm outputs, the price of chemical fertilizer differs significantly among the Central Asian republics (see Table 5.24). Both Uzbekistan and Turk-

TABLE 5.20 Production of nitrogenous fertilizers, 1992-96 (thousands of tons)

	1992	1994	1996
Kazakhstan	64	96	87
Kyrgyz Republic	n.a.	n.a.	n.a.
Tajikistan	48	10	10
Turkmenistan	79	102	110
Uzbekistan	1,018	673	818

SOURCES: FAO various years; and World Bank 1993, 1994, 1995a, 1995b, 1996, and 1998a.

NOTE: n.a. indicates not available.

TABLE 5.21 Exports of nitrogenous fertilizers, 1992-96 (thousands of tons)

	1992	1994	1996
Kazakhstan	24	16	51
Kyrgyz Republic	n.a.	n.a.	n.a.
Tajikistan	10	n.a.	n.a.
Turkmenistan	n.a.	n.a.	n.a.
Uzbekistan	239	53	220

SOURCES: FAO various years; and World Bank 1993, 1994, and 1997.

NOTE: n.a. indicates not available.

menistan continue to provide fertilizers to producers at below-market prices. These price differentials create significant incentives for producers in Uzbekistan to sell subsidized fertilizers to producers in neighboring Kazakhstan and the Kyrgyz Republic, where fertilizer prices are much higher. This is highly inefficient because the difference accrues to the individual seller of the resource—who is not making any productive contribution to the economy—rather than to the producer or to the state.

Evaluations of the fertilizer sector in those Central Asian republics that have privatized this industry have generally concluded that this sector is operating at suboptimal efficiency (World Bank 1994). Numerous structural impediments preclude the entry of new agents into the market while complicated regulatory frameworks render trading in this good cumbersome. Evidence from other regions suggest that the state will have to provide support services—in the form of credit, market information, and infrastructure—to bolster private sector activity.

TABLE 5.22 Production of phosphate fertilizers, 1992-96 (thousands of tons)

	1992	1994	1996
Kazakhstan	400	70	175
Kyrgyz Republic	n.a.	n.a.	n.a.
Tajikistan	n.a.	n.a.	n.a.
Turkmenistan	130	13	20
Uzbekistan	310	129	194

SOURCES: FAO various years; and World Bank 1994, 1996, and 1997.

NOTE: n.a. indicates not available.

TABLE 5.23 Imports of potash fertilizers, 1992-96 (thousands of tons)

	1992	1994	1996
Kazakhstan	n.a.	n.a.	6
Kyrgyz Republic	5	5	5
Tajikistan	8	5	5
Turkmenistan	10	10	11
Uzbekistan	60	50	75

SOURCES: FAO various years; and World Bank 1993, 1994, 1996, and 1998a.

NOTE: n.a. indicates not available.

TABLE 5.24 Cost per kilogram of nutrient (U.S. dollars)

	Nitrogen (N)	Phosphorous (P)	Potassium (K)
Kazakhstan	.5	1.5	.16
Kyrgyz Republic	.5	1.5	.16
Tajikistan	.5	1.0	.15
Turkmenistan	.12	1.0	.04
Uzbekistan	.25	0.5	.07

SOURCE: EU-TACIS 1995.

Institutional Constraints

The transition from a centrally controlled system to a market system requires not only the abolition of the price controls and subsidies inherent in the old system, but also the building of new institutions. The new institutions should be adequate for the reformed structure of property rights affecting factor markets (particularly land, water, and credit) and the emergence of a private distribution system. Unless these institutions are set in place, the movement toward the market will face several bottlenecks and risk failure. Moreover, in the transition toward the market, there is sometimes the tendency to eliminate support even to those activities that require an active role of the state because of their public goods nature. In the following paragraphs, we focus on some of these issues, such as those related to water use, seed multiplication, and the research system.

Water. Central Asia is marked by a range of agroclimatic conditions. Agriculture in the northern regions of Kazakhstan is characterized by extensive rainfed grain and livestock production. Low rainfall and highly variable weather create a high-risk production environment. The south of Central Asia is mainly arid or semi-arid so that nearly all agriculture depends on irrigation.

Table 5.25 shows the percentage of irrigated land in the five Central Asian republics in relation to Iran, Pakistan, and Turkey, countries with similar growing conditions. It makes clear Central Asia's heavy reliance on irrigation and the corresponding importance of water management to its agricultural policy.

The Central Asian republics inherited centralized irrigation structures in place since the Soviet era. With the dissolution of large state farms into smaller holdings, these structures are proving inadequate. The physical infrastructure, built to accommodate large landholdings, is inappropriate for reaching individual farms. Moreover, since the management system remains centralized, smallholders are unable to adjust the delivery of water to suit their production needs. Effective reform therefore requires that water management systems be decentralized in terms of both physical infrastructure and management.

TABLE 5.25 Land use

	Arable land as a percent of total land	Hectares of arable land per capita	Irrigated land as a percent of arable land
Kazakhstan	12.0	2.08	7
Kyrgyz Rep.	7.0	0.32	78
Tajikistan	5.4	0.15	84
Turkmenistan	3.5	0.37	88
Uzbekistan	11.0	0.20	89
Turkey	32.0	0.46	15
Iran	11.0	0.28	39
Pakistan	27.0	0.17	80

SOURCES: World Bank 1999; and FAO various years

Evidence suggests considerable wastage through grossly inefficient practices. Table 5.26 provides estimates of the efficiency of water use for cotton production in the Central Asian republics compared with other agricultural regions. These estimates suggest that one quarter of water used in Central Asia is lost through the interfarm water delivery system. Research could help determine the more precise form new systems should take.

Seed sector and research institutions. Prior to 1991, the Soviet Union had developed an impressive infrastructure of agricultural research institutes and support services. Unfortunately, since the breakup, these institutes have suffered from reduced investment. In Turkmenistan and Uzbekistan, the state has maintained a large presence in the seed sector, and the industry is experiencing fewer problems. Producers in these countries generally do not report difficulty in procuring seed for major crops. In contrast, in Kazakhstan, the Kyrgyz Republic, and Tajikistan, the seed industry is in disarray. In these regions it has become common for producers to recycle stored grain over several growing seasons, which necessitates heavier seeding rates as the germination rate and productivity of this seed falls over time.

A wide range of research has demonstrated that the return to investment in agricultural research is very high. As such, policymakers in the region may wish to reevaluate recent policies toward investment in agricultural research and to implement structures through which agricultural research can operate effectively in a reformed economic environment. Even in a market economy the state should play a proactive role in agricultural research and development. Most observers argue that its role should be limited to the provision of public goods: those that stand to benefit society as a whole but which the private sector does not have the incentive to provide.

TABLE 5.26 Cotton production and water use in Central Asia and other selected areas

	Seed cotton yield, 1996-98 average (tons per hectare)	Water use (thousands of cubic meters)	Kg of seed cotton/ thousand cubic meters of water
Kazakhstan	1.79	5.8	309
Kyrgyz Republic	2.28	9.9	230
Tajikistan	1.60	12.8	125
Turkmenistan	1.51	5.9	256
Uzbekistan	2.40	8.8	273
	Seed cotton yield, 1996-98 average (tons per hectare)	Water use (thousands of cubic meters)	Kg of seed cotton/ thousand cubic meters of water
California	4.48	9.2	487
Australia	4.27	7.0	610
Greece	2.77	2.7	1,027
Syria	3.52	7.4	462

SOURCES: Calculated from Lerman, Garcia-Garcia, and Wichelns 1996; EU-TACIS 1995; World Bank 1991, 1993, 1994, 1995a, 1995b, 1996, 1997, and 1998a; and FAO various years.

Private companies will not fund this sort of research because they stand to bear all of the cost while gains tend to be more diffuse. The development of an improved seed variety is an example of a public good, while the private sector can typically be relied upon to multiply the seeds. However, the state may still need to play a role in supporting these nascent industries until they can be self-supporting. Donor agencies have sought to foster the renovation of this industry through a variety of programs. Currently they are funding projects in the Kyrgyz Republic and in Uzbekistan to help develop private cotton seed multiplication farms.

Research to Improve Reform of Agricultural Input and Output Markets

The economic and political environment in Central Asia region during the past decade has changed considerably while the level of resources allocated to research has been minimal. Because of agriculture's importance to the region's economy, and because agriculture is facing significant problems, there is a great need for research that could help formulate policies fostering strong and sustainable development. Following is a preliminary discussion of some policy issues future research should address. We make no effort to prioritize these needs because of insufficient information and because they will vary according to the specific circumstances in each country.

Efficiency in Input and Output Markets

The movement toward market reform is often perceived as a reaction to the unsatisfactory performance of the centrally planned system. State organizations involved in the production and distribution of agricultural outputs and modern inputs have come under attack because of the high inefficiencies involved in their operations. Output price controls have resulted in a lack of incentives for producers, and mounting subsidies to the use and production of modern inputs have become unsustainable in most developing countries, necessitating market reforms. The concern for improving the efficiency of the current system suggests the set of general policy questions to which research could provide some useful answers. Some key questions for policy research in this area follow:

- How can the distribution costs of the input and output delivery system be reduced?
- What is the balance between public and private roles?
- Which institutional reforms should complement market reforms in order to improve the performance of the input and output delivery system?

Emergence of an Efficient Private Marketing System

Even if market reforms are undertaken, a thriving and efficient private sector will not necessarily develop and engage in those functions previously performed by the public sector. In the presence of market failures and infrastructure bottlenecks, the effects of market reforms on agricultural marketing may be adverse. Sometimes, governments give exclusive rights to one major private domestic or foreign firm, limiting farmers' access to technology, and institutionalizing barriers to market entry. In other cases, imports are limited to particular brands, restricting farmers' access to a broader range of choices available in the international markets. The private sector may not have an incentive to participate in the marketing of agricultural inputs because of thin markets or the lack of credit institutions. This set of issues leads to a set of questions related to what the government could do to remove the obstacles to the development of an efficient private sector.

- Is the removal of subsidies, trade restrictions, and marketing state organizations leading to the development of an efficient private marketing system?
- What can the government do to promote the emergence of a thriving marketing system?

Effects of Reforms on Farmers

If market reforms are introduced, the system may become more efficient. Reform, however, will have differential impact on various kinds of producers. It is thus important to understand the effect of reform on different social groups, especially "winners" and "losers." With respect to winners, it is crucial to understand how certain social groups—such as large farmers—are able both to influence the pattern of market reform and to benefit from such reform. With respect to losers, it is important to analyze how other social groups—such as small farmers—are able to cope with the possibly deleterious side-effects of market reforms. For instance, input market reforms may lead to higher production costs, which will have a negative effect on the incomes of small producers. This case raises three questions:

- What are the effects of input and output market reforms on the income of different farmers?
- What is the resulting small farmers' access and use of modern inputs?
- What is the effect on the level and sustainability of agricultural production?

Sequencing of Reforms

Experience shows that one major difficulty in the reform process consists in specifying the appropriate sequencing of policy measures. The success of input market liberalization closely depends on the right policies affecting output markets and credit institutions. Removal of distortions in one market may have undesirable results when other related markets continue to be distorted. For instance, the withdrawal of trade restrictions on irrigation equipment in Bangladesh was accompanied by strong growth in production, with the procurement system at support prices still in place. That created incentives to over-supply rice with a subsequent precipitous fall in prices when, two years later, the government decided to basically eliminate domestic procurement. In the case of Kazakhstan, the withdrawal of input subsidies prior to liberalization of output prices had adverse effects on production.

A gradual approach to reform is often advocated because of the differential speed of price response in input and output markets. Shock therapies are perceived as compromising the effectiveness of the proposed changes, especially in those cases where heavily controlled systems have been in place for long periods of time. In Mexico, for example, the dramatic changes carried out in the past few years are deeply altering the institutional relations built over a very long period of history. It is difficult to understand how political, social, and economic relations could be reformed in a short period of time without inflicting tremendous stress on the system. These considerations suggest the following policy questions.

- How can input markets be reformed successfully in conjunction with output markets?
- What is the appropriate sequencing of these linked reforms?
- Is a gradual approach more effective than shock therapy in the transition to a new balance between the public and private sectors in input markets?

Comparative Advantage of Grains versus Cash Crops and Livestock

Certain republics, notably Turkmenistan and Uzbekistan, have enacted policies to ensure more wheat production in place of cotton, in order to become more self-sufficient in cereal production. However, the merit of such a policy needs more rigorous evaluation. Its success will depend on whether the countries' resource endowments are more ideally suited to cotton or grain. If they are more suited to cotton production, the countries may be best advised to use cotton receipts to import grain. Alternatively, they might benefit more from devoting even more space to grain or to other activities. This question should be explored in depth in each of the republics to determine where their comparative advantage lies. Similarly, the huge potential for livestock and meat products in Central Asia has received increased attention. The key question in this case is to understand to what extent the livestock sector potential can be translated into comparative and competitive advantage in the region.

Regional and International Trade

A related issue is the effect of transport costs on producer and export prices. In the case of Kazakhstan such costs can be considerable. The development of an infrastructural network—and particularly one that ensures access to port facilities—can thus play a role in moderating prices and increasing overall efficiency. However, the question of relative costs and benefits to be derived from different sorts of infrastructure must be explored. Moreover, the relative benefits of various overseas outlets should be examined to determine the most cost-effective destination for produce.

Reforms in Water Management

Research to evaluate alternative policy options in this area would be useful. A great deal of benefit could be derived from improved water use patterns at all levels of the water delivery system. In addition, it would be valuable to investigate alternative mechanisms by which appropriate water management structures or water use associations could be developed at the producer level. Improvements in water policy could yield not only higher production, but substantial environmental and social benefits as well.

Land Tenure and Farm Size

The dismantling of central planning has led to the emergence of several farm ownership models over the past several years. However, it is not entirely clear which ownership structure is the most efficient and most conducive to the welfare of small-holder farmers. It would be useful to know more about the effects of different ownership structures on agricultural development as well as which types are best suited to changing economic circumstances.

Conclusion

Following the dissolution of the Soviet Union, the Central Asian republics implemented significant policy reforms to respond to the changed economic environment. They have achieved progress in some areas but still face substantial challenges to embark on a strong and sustained growth path. The discussion in this paper suggests agriculture can be a significant source of growth. However, research on the agricultural and natural resource sectors of the region is needed to provide policymakers with informed analyses and evaluations of policy alternatives. Research in these sectors in other countries has proven very useful in improving rural welfare. In Central Asia too, such research could prove to be of significant benefit in fostering sustainable and equitable growth.

References

Babu, S. C., and A. Tashmatov. 1999. Attaining food security in Central Asia-Emerging issues and challenges for policy research. Food Policy 24: 357-362.

Carana Corp. 1998. Meat and dairy systems analysis for the Republic of Kazakhstan. Prepared for USAID Central Asia Regional Mission, Almaty.

Chabot, P., and S. Kyle. 1998. Agricultural land use in northwest Uzbekistan: A linear programming model for mapping producer incentive. Working Paper #98-13. Ithaca, N.Y.: Cornell University Department of Agricultural, Resource, and Managerial Economics.

Craumer, P. 1995. Rural and agricultural development in Uzbekistan. London: The Royal Institute of International Affairs.

EU-TACIS (European Union-Technical Assistance to the Commonwealth of Independent States). 1995. Preliminary report: Irrigated crop production systems. Water and resources management and agricultural production in the Central Asian republics (WARMAP) project report. Tashkent

_____. 1997. Water use and farm management survey. Water and resources management and agricultural production in the Central Asian republics (WARMAP) project report. Tashkent.

FAO (Food and Agricultural Organization of the United Nations). Various years. FAO production yearbook and agricultural statistics. Rome.

Griffin, K. 1996. Social policy and economic transformation in Uzbekistan. Geneva: United Nations International Labor Organization.

Hutchens, A., and P. C. Mann. 1998. Review of water pricing policies, institutions and practices in Central Asia, Environmental Policy and Technology Project. Prepared for USAID Central Asia Regional Mission.

International Cotton Advisory Committee. 1993. The world cotton market: Prospects for the nineties. Washington, D.C.

IMF (International Monetary Fund). 1998a. Turkmenistan: Recent economic developments. Washington, D.C.

_____. 1998b. Republic of Kazakhstan: Recent economic developments. Washington, D.C.

_____. 1998c. Republic of Tajikistan: Recent economic developments. Washington, D.C.

_____. 1998d. Republic of Uzbekistan: Recent economic developments. Washington, D.C.

_____. 1999. Krygyz Republic: Recent economic developments. Washington, D.C.

Johnson, S., and B. Islamov. 1991. Property rights and economic reform in Uzbekistan. Wider Working Paper 90. Helsinki: World Institute for Development Economics Research.

Krueger, A., M. Schiff, and A. Valdes. 1992. The political economy of agricultural price intervention in Latin America. San Franciso: ICS Press.

Kyle, S., and P. Chabot. 1997. Agriculture in the Republic of Karakalpasktan and Khorezm Oblast of Uzbekistan. Working Paper #97-13. Ithaca, N.Y.: Cornell University Department of Agricultural, Resource, and Managerial Economics.

Lerman, Z., J. Garcia-Garcia, and D. Wichelns. 1996. Land and water policies in Uzbekistan. Post-Soviet Geography and Economics 37 (3).

Michaelopoulos, C., and D. G. Tarr. 1994. Trade in the new independent states. Washington, D.C.: The World Bank.

Micklin, P. 1991. The water management crisis in Soviet Central Asia. The Carl Back Papers. University of Pittsburgh Center for Russian and East European Studies, Pittsburgh, Penn.

Micklin, P., and W. Williams. 1996. The Aral Sea Basin. New York: Springer Publishers in cooperation with NATO Scientific Affairs Division.

Smith, D. 1994. Change and variability in climate and ecosystem decline in Aral Sea Basin deltas. Post-Soviet Geography and Economics 35 (3).

Spoor, M. 1993. Transition to market economies in former Soviet Central Asia: Dependency, cotton and water. Working Paper 160. Le Hague, The Netherlands: Institute of Social Studies.

_____. 1995. Agrarian transition in former Soviet Central Asia: A comparative study of Uzbekistan and Kyrgystan. Working Paper 202. Le Hague, The Netherlands: Institute of Social Studies.

_____. 1997. Agrarian transition in the former Soviet Union: The case of Central Asia, the market panacea. London: Intermediate Technology Publications.

Taube, G., and J. Zettlemeyer. 1998. Output decline and recovery in Uzbekistan: Past performance and future prospects. Working Paper 98/132. Washington, D.C.: International Monetary Fund.

USDA (U.S. Department of Agriculture). Various years. Foreign agricultural statistics. Washington, D.C.

Wilson, R. T. 1997. Livestock pastures and the environment in the Kyrgyz Republic, Central Asia. Mountain Research and Development 17 (1).

World Bank. 1991. World cotton production: Prospects to 2005. Working Paper 287. Washington, D.C.

______. 1993. Uzbekistan: An agenda for economic reform. Washington, D.C.

______. 1994. The Kyrgyz Republic: Policy note on competitive input marketing in the crops sector. In Kazakhstan: Agricultural sector review. Washington, D.C.

______. 1995a. Krygyz Republic: Agricultural sector review. Washington, D.C.

______. 1995b. Republic of Uzbekistan cotton sub-sector improvement project. Washington, D.C.

______. 1996. Turkmenistan: Review of the agrarian sector. Washington, D.C.

______. 1997. Uzbekistan farm restructuring and development program feasibility study. Washington, D.C.

______. 1998a. Krygyz Republic agricultural policy review. Washington, D.C.

______. 1998b. World development indicators. Washington, D.C.

______. 1999. World development indicators. Washington, D.C. CD-ROM.

6 Rangelands and Livestock Systems in Central Asia: Policy Research Needs

PETER ORAM

Natural grazings comprise the largest area of usable agricultural land in Central Asia, supporting a pastoral society with around 15 million cattle and 40 million sheep. However, as in other comparable geographic regions, the pastoral societies have received low priority from governments, their lands have been sequestered for other uses, and they are under threat from past mismanagement and resource degradation. Addressing these issues represents a major challenge to the newly independent governments of the region.

Drawing on recent reports from Central Asia and other geographic regions, this chapter identifies four major areas for action and suggests possible solutions. The four areas are these:

- *Resource degradation.* This problem must be examined from many perspectives: its causes and effects within and across land use systems (particularly rangelands and forests), the need to understand cross-system interactions, and the importance of an integrated and participatory approach to better management of upland watersheds.
- *Water.* This is the crucial limiting factor for pastoral systems. Key issues include availability, access rights, resource management by and for communities, and planning future needs. Water for irrigation is an intercountry priority in Central Asia; it should also be treated as such for rangelands, possibly through a regional commission for which the Interstate Commission for Water Coordination (ICWC) provides a precedent.
- *Land tenure.* The recent historical transformation from central control and exploitation of resources to a more open regime in Central Asia raises numerous questions as to the optimum system of stewardship and rights to use or ownership of the rangelands, for example through collective, communal, or private tenure. The paper draws on extensive recent research in West Africa and North Africa to identify institutional forms that can provide secure tenure, facilitate productive investment, and provide incentives to protect the environment.
- *Drought management.* Drought is a permanent threat to rangeland people and their animals; severe droughts can have devastating social, economic, and environmental consequences, but government responses in other regions have generally been reactive rather than pro-active. Those experiences indicate the need to develop a comprehensive national drought man-

agement and investment strategy with a solid infrastructure and institutional underpinning transcending the public sector; and the chapter suggests what its components might be.

Background to Rangelands and Livestock Issues

Rangelands in Central Asia

The Central Asian republics, which have an estimated pasture area of 9 million to 187 million hectares (Table 6.1), have not been immune from problems. They form part of an even larger region of arid lands, including parts of Russia, Mongolia, and China. Historically, the main land use in this vast area was extensive migratory livestock production, without rigidly defined state lines. Animals were moved within and between ecoregions to take advantage of changes in vegetation from summer to winter and to obtain access to water, which is usually the major physical constraint to animal husbandry in arid areas.

Currently, the independent governments of the Central Asian republics are struggling to rebuild their economies and reorient their markets. At the same time, they have to repair the environmental damage resulting from the overextension of irrigated area, inefficient use of water and agrochemicals for irrigated crops (especially cotton), the ploughing up of natural grazings for rainfed crop production in a predominantly arid climate, and the escalation of ruminant livestock numbers in pursuit of centrally planned production quotas. This is a Herculean task involving significant shifts in the role of the state away from a command economy—especially with respect to restoring those elements of the land tenure and resource management systems which worked in the past—and toward establishing or strengthening resource monitoring and quality control systems which are acceptable to local communities. Restoring sustainability to resource use should be a first priority, even if it appears to conflict with short-run production objectives.

To some extent this task may be facilitated by the decline in ruminant livestock numbers which has occurred in most Central Asian countries since 1990 (Table 6.2). This decline may temporarily ease pressures on the natural grazings, while improved environmentally sound systems of management are developed with greater decentralization of responsibility to local authorities, land users, and flock owners. However, at the same time, the decline in livestock numbers is creating social and economic problems as a result of deteriorating rural incomes along with declining transport and other rural services which affect access to markets, feed, and other production inputs, as well as to water.

Threats to the Rangelands

In many countries the natural pastures are the last frontier of uncultivated land. Large areas have been converted over time to irrigated or rainfed cropland as well

TABLE 6.1 Land use in Central Asian countries, 1994-95 (thousands of hectares)

Country	Total land	Crop-land	% of total land	Pastures[a]	% of total land	% in moun-tains	Forests[a]	% of total land	Other land[b]	% of total land	Irrigated land	% of crop-land
Kazakhstan	267,073	34,978	13.1	186,823	70.0	11	9,600	3.6	35,672	13.4	2,300	6.6
Kyrgyz Republic	19,180	1,420	7.4	9,000	46.9	95	730	3.8	8,030	41.9	1,050	73.9
Tajikistan	14,060	860	2.1	3,550	25.2	15	537	3.8	9,113	64.8	718	74.8
Turkmenistan	46,993	1,480	3.1	30,000	63.8	90	4,000	8.5	11,513	24.5	1,300	87.8
Uzbekistan	41,424	4,500	10.9	20,800	50.2	11	1,300	3.1	14,824	35.8	4,000	88.9

SOURCE: FAO 1996.

NOTES: FAO data are not available for some parameters after 1994; in those cases 1994 data are cited.

[a] FAO data for pastures and forests are sometimes unclear because of different definitions used by reporting countries.

[b] Land unsuitable for agriculture (mountains, deserts, wetlands, etc.), urban areas, roads, etc.

TABLE 6.2 Numbers of ruminant livestock in Central Asia, 1992 and 1996 (thousands of head)

	Cattle			Sheep			Goats		
Country	1992	1996	% change 1992-96	1992	1996	% change 1992-96	1992	1996	% change 1992-96
Kazakhstan	9,084	6,859	-25.0	33,908	18,725	-44.7	692	799	+15.4
Kyrgyz Republic	1,190	869	-27.0	9,225	4,075	-55.8	300	200	-33.3
Tajikistan	1,222	1,147	-6.2	2,172	1,783	-18.1	840	710	-15.5
Turkmenistan	777	1,199	+54.3	5,380	6,150	+14.4	220	424	+92.7
Uzbekistan	5,113	5,204	+1.8	8,275	8,352	+0.9	918	970	+5.7

SOURCE: FAO 1997.

as to urban and industrial use, particularly since World War II, under pressure of rapid population growth and with the expansion of mechanized farming. While this has reduced the land available to graziers, it has not necessarily induced a concomitant decline in stocking rates. Rising national populations and income growth have raised demand for animal products; and autarchic policies of food self-sufficiency have provided subsidies, cheap fuel, and other incentives to keep more animals.

One consequence in a number of developing countries has been an increase in the grazing of forested areas, often in upland watersheds.[1] This, as well as increased livestock populations on traditional pasture lands, has accelerated erosion and land degradation. According to one analysis, overgrazing is the single most important source of degradation worldwide, representing 35 percent of the overall global total (World Resources Institute 1992).

The most obvious adverse effect of resource degradation is the in situ impact on the state of the grazings, creating a downward spiral of declining vegetative quality and quantity. In addition, resource degradation—especially in arid regions with fragile soils and upland areas with steep slopes—can generate serious off-site externalities, causing landslides, sandstorms, and flooding, with damage to roads, houses, and power systems and silting of rivers, reservoirs, and irrigation canals. These effects are a pervasive problem reported by virtually every ecoregion and a variety of land ownership and management regimes. Despite this evidence, deforestation rather than overgrazing is often considered the main causal agent, possibly because it is more obvious and easier to measure than range degradation.

Analogies between Rangelands in Central Asia and Other Geographical Regions

Ecologically, there are affinities between Central Asia; the western United States; western Argentina; southern, central, and western Australia; and West Asia-North Africa (WANA). However, only parts of the Rocky Mountain and prairie states of the United States, and a few West Asian countries have such cold winters as Central Asia. Also, state ownership of grazing lands is more common in the United States (with leasehold rights to users) and WANA, than in Australia or Argentina.

Historically, it is hard to find a situation parallel to that in Central Asia, where five countries—previously part of a huge, centrally planned, and authoritarian empire—suddenly became independent in a single year. The same cannot be said of any of the ecologically analogous countries of the Americas or Australia; and the dissolution of the Ottoman Empire was much more gradual, with Turkey remaining a major sovereign state. Perhaps the closest comparison

---

[1] An estimated 55 percent of state forest lands in the Kyrgyz Republic is leased by the leskhoges (collectives) to herders, including people from neighboring countries.

to Central Asia is the progressive decolonization and independence of the WANA countries from British, French, and Italian rule between the end of World War II and the 1970s. Although as a group those countries are probably more diverse politically, economically, and ecologically than the Central Asian countries, many of their problems are quite similar. Hence the approaches being taken to rangeland and livestock management there may be relevant to Central Asian countries.

Rangeland and Livestock Systems in Central Asia and WANA: Common Problems and Their Implications for Policy

Rangeland and livestock systems in both Central Asia and WANA are mainly confined to the lower rainfall and often more remote ecozones (now often defined as "less favored" areas), and thus have tended to receive lower priority from governments than the more favorable rainfed and irrigated areas. This has often been reflected in lower expenditures on roads and social infrastructure (health, education, rural water, and electricity supply) and in more limited support for research and technology transfer efforts. Thus the inhabitants of these areas are often poverty stricken, and they are a main source of out-migration to urban areas or overseas.

Livestock systems in both these regions are extensive, with small ruminants (especially sheep) as the mainstay of the system and often the primary source of family incomes, but with goats and cattle also playing a significant role. Comparing the numbers of animals in the two geographic regions indicates that cattle are of somewhat greater importance in Central Asia than in WANA, and that animal husbandry is more extensive (more land per head of livestock) in Central Asia. This is in line with the much larger share of land used for grazing in Central Asia, with under 100 millimeters of annual precipitation, and the extent of migratory herding (Table 6.2).

Despite these interregional differences, many common problems need to be addressed by policymakers, some of which have multiple roots and interactions throughout the agricultural sector and the economy, such as:

- Resource degradation in irrigated and rainfed cropland, rangelands, and forests
- Water scarcity, wasteful use, and intersectoral competition for water
- Land tenure insecurity, state interventions, and their impact on sustainable land use
- Low government investment in less favored areas
- Drought frequency and government reactions, for example, feed, fuel, and transport subsidies; government shrub and cactus plantations on tribal lands as emergency feed reserves; bans on cultivation in Livestock Rearing Areas (LRAs) below 200 millimeters annual precipitation; well-sinking, etc.
- The declining contribution of rangelands to animal feed supply, and overall feed deficits

- The transition from an autarchic to a market-oriented economy, and the impact of structural adjustment programs on government policies and the rural economy in the LRAs

Resource Degradation

Resource degradation is a pervasive problem in both Central Asia and WANA, and it affects virtually all forms of land use, including irrigated and rainfed arable farming, steppe and upland grazing, and forests and watersheds, often with backward and forward linkages between the different land use systems. For example, converting semi-arid rangelands into cropland puts pressure on the remaining grazings and may also lead to invasion of upland pastures and forests by more livestock, which in turn can increase soil and water erosion from the upland watersheds, leading to landslides, flooding, and other off-site problems. These interactions and damaging externalities were reported for virtually all ecozones in Sub-Saharan Africa, WANA, South and Southeast Asia, China, and Latin America at an international meeting convened by the International Food Policy Research Institute (IFPRI) in 1994 to discuss ecoregional characterization as a tool for identifying key natural resource problems and their implications for economic growth and sustainable resource management.

Issues Affecting Rangeland Degradation

For the WANA region, the International Center for Agricultural Research in the Dry Areas (ICARDA) undertook a cross-country comparative analysis of the current situation and future outlook for ruminant animal feed supply. The results show that, in a majority of WANA countries, the contribution of the natural grazings to total supply has declined significantly and in some cases dramatically, resulting in a shift away from dependence on pastures toward supplementary feed and crop residues (Nordblom and Shomo 1995). The outlook to 2020 for most WANA countries is one of contracting domestically produced feed supply, escalating feed imports, and a decline in per capita consumption of animal products, especially sheep meat for the poorer countries and people. Several factors are responsible for this serious situation:

- The conversion of grazing lands to crop production, in particular the expansion of mechanized barley cultivation onto the range;
- The elimination of annual fallows and fallow grazings in the low rainfall crop rotations in favor of continuous crop production. Between 1975 and 1990 the fallow area in WANA fell by about 10 million hectares from 41.40 percent to 29.16 percent of the total arable area (Oram and Belaid 1990);
- The influence of government subsidy and other policies, including cheap fuel, which facilitated the mechanized cultivation of pastures in low rainfall

areas and enabled better-off pastoralists with trucks to move feed and water to flocks parked on the rangelands for much longer periods than in the past, thus jeopardizing the recovery of the natural vegetation after the end of the normal grazing period;

- Untargeted provision of feed grains to flock owners at subsidized prices or as drought relief handouts. These became institutionalized in some countries and eventually a burden on their treasuries (Oram 1997). Their abrupt cessation in Jordan as a condition of structural adjustment, which coincided with a serious drought year and an embargo on cross-border sheep exports to Saudi Arabia, led to the abrupt slaughter of an estimated 30 percent of the national flock;
- Price supports for domestic production of cereals at levels well in excess of world prices and also for meat production, which simultaneously provided incentives for "sod-busting" of marginal lands by private entrepreneurs and for flock owners to keep more animals on a shrinking area of rangeland;
- Expansion of irrigated lands, mainly through government projects and usually at the expense of pastures in previously uncultivated low rainfall areas. However, in some WANA countries private ground water exploitation for crop production is also increasing, mainly as a result of subsidized fuel or electricity for pumps. This expansion is not only taking over some grazing land but, more seriously, because of the failure to regulate the numbers of wells and types of pumps, it is leading to dramatic lowering of the ground water level; and
- The nationalization of the natural grazings by governments, with restraints on nomadic systems of land use and the settlement of nomadic people, leaving a vacuum in the control and management of the resource which state governments and state-dependent cooperatives have had difficulty filling effectively, and which has led to open access and competition for use.[2]

Contrary to these trends, it is now widely accepted that (far from being a cause of degradation) migratory systems of ruminant livestock husbandry along agreed upon territorial lines of mobility are probably the most efficient mode of land use in the water scarce, risky, and drought-prone low-rainfall areas which characterize most of WANA and Central Asia. In the context of that harsh and uncertain environment, the concept of imposing centrally planned production quotas on range-based ruminant livestock, as was done in Central Asia, does not seem economically or ecologically sound and has probably been a significant contributory factor to range degradation there.

Rather than trying to maintain central control, the challenge to governments of the region is to reach an effective agreement with the flockmasters (1) with

[2]According to one study, all the Bedouins in Syria are now transhumants with bases in villages, and there is virtually no pure nomadism (Masri 1991).

respect to grazing fees, their rights to use the rangelands, and how to manage the rangelands sustainably; (2) to provide them with access to the water, knowledge, inputs, credit, and markets they need; and (3) to establish a resource monitoring system to provide early warning of drought or of further degradation and its causes, with power to impose penalties for deliberate mismanagement of the resource if consultation with the users fails to redress the problem.

The Impact of Grazing on Forests

A related problem which policymakers must address is that of grazing in forested areas, which are also vulnerable to damage from fire, illegal wood cutting, pests, and diseases. According to a recent assessment, the forests in Kazakhstan, which has the largest area of forests in Central Asia, declined by nearly 2 percent per year between 1990 and 1995 mainly because of over cutting and grazing; and 10 percent of its coniferous forests were lost to fire in 1997 because of a lack of surveillance. Rapid deforestation is also reported for the Kyrgyz Republic, possibly reducing the forest cover by 50 percent since 1945; this is attributed to heavy sheep grazing and to local demands for fuel wood and construction (Government of Kyrgyzstan 1995). Similar problems are reported for WANA countries, with serious threats to wildlife and biodiversity (Taimeh 1997).

Forests and woodlands occur both in the lowland desert areas (especially the Saksaul forests in Kazakhstan) and in the mountain areas of the region. Although they total in aggregate only about 6.5 percent of land area in Central Asia (as in WANA), they contain a wealth of species biodiversity, including white saksaul, spruce, pistachio, almond, juniper, walnut, apple, and maple. Protected areas have been established in Kazakhstan, the Kyrgyz Republic, Tajikistan, and Turkmenistan, but the countries are having trouble maintaining them.

In addition to providing grazings, the Central Asian forests play an important role in protecting the Aral Sea watershed, reducing siltation of dams and helping control desertification. They are a valuable reservoir of biodiversity and a potential source of revenue from tourism.

However, because they do not contribute substantially to gross domestic product, financial support from the states to the various government organizations responsible for forest management, monitoring, and administration is generally inadequate to ensure sound land use, prevent abuse, and oversee protected areas. This problem is related to the difficulty of valuing the indirect use and nonuse benefits of forest resources, which may well exceed the direct value components such as timber, normally used as the basis for their economic assessment.

It is not the purpose of this chapter to discuss in depth the range of forest policies being adopted by governments in Central Asia or in neighboring countries with analogous ecological situations (Albania, Turkey, and the Caucasus, for example). These policies differ widely, and in several countries including Kazakhstan, the Kyrgyz Republic, and Tajikistan, there have been recent changes in institutional mechanisms. However, participatory watershed management—

involving local communities, nongovernmental organizations (NGOs), and provincial rural service agencies collaborating on problem identification, formulation of development plans, and implementation—is reported to have worked well in Turkey. It has been suggested that linking forest management to collaborative natural resource management and rural development would be an appropriate strategy for other countries of the Caucasus and Central Asia as well. An integrated approach to sustainable land management in upland watersheds is essential.

It is noteworthy that several of these countries are considering arrangements to decentralize forest management to local authorities and/or to communities, and that NGOs are playing a more active partnership role in forestry and management of biodiversity, building on local traditions and linkages. One important goal of these community oriented developments should be reaching agreement with flock owners on grazing rights, stocking levels, and grazing periods with respect to forests and woodlands. Charges should be set to limit the time and numbers of stock so as to preserve the long-term integrity of the resource. Ecotourist potential exists in some forest and mountain areas, and its development should also be planned with the local communities, both as stakeholders and stewards of the resource.

Water Resources for Rangeland Development

Two main aspects of water policy are relevant to the sustainable use of rangelands: access to water for people and their flocks, and irrigation development and irrigated land use. Water is fundamental to the survival of people and animals in arid rangelands. The ability to move flocks within and across national borders to obtain access to water has been a driving force for traditional migratory systems of range management. Those systems, in turn, have helped maintain the precarious balance between stock numbers and the natural vegetation, and provided some insurance against severe droughts.

Water for livestock in the LRAs can come from a variety of sources, including irrigation systems, rivers, streams, wells, and water harvesting and storage systems (ponds, tanks, cisterns, and large underground constructions, etc., as in Saudi Arabia, Iran, Baluchistan, and the Yemen). Common problems include severe overgrazing around wells, particularly in settlement areas; overpumping and poor siting of wells, which lower and dry out the groundwater table; and failure to maintain water harvesting and storage systems. Rights of ownership and access to water vary from communal (based on customary traditions) to private (tube wells), to municipal (settlements), to state (irrigation projects). Well construction may be subsidized or financed by the state, with responsibility for maintenance and pumping charged to the owners/users.

Factors which have significantly affected access to and availability of water to rangeland users in recent years include:

- creation of newly independent states and, in some cases, the establishment of territorial borders where none existed previously;
- settlement of nomadic people and proscription of the tribal rights to unrestricted customary use of rangeland resources. This has led to heavy concentrations of people and animals around water points in settlement areas, with severe range degradation, while maintenance of water storage systems in more remote areas has often been neglected;
- nationalization of rangelands, exacerbating existing uncertainties and conflicts over property rights, and in some cases opening the rangeland to competition for land and water resources among the herders, and between them and opportunistic private cultivators;
- failure of governments to establish effective links with users to ensure sustainable management and stewardship of the resource;
- emphasis on increasing the numbers of livestock to meet production quotas rather than raising productivity per head, while at the same time sequestering land and water resources to expand irrigated area and rainfed crop production; and
- wasteful and inefficient use of irrigation water and concentration of land use on crops such as cotton with a high water requirement and susceptibility to pests, causing pesticide pollution, salinization of soils and surface and ground water, and almost total extraction of water from the two major river systems in Central Asia, severely restricting access to river water and to irrigated land for grazing crop residues by ruminant livestock.

Another consideration is the economics of producing more grain or forage legumes in irrigated or rainfed crop rotations for sale to herders in the rangelands or for feeding to livestock *in situ*. This practice could help maintain fertility on irrigated land to offset current problems of fertilizer availability, and reduce national feed deficits. It has long been the practice in Egypt, another country where cotton and wheat are important components of the irrigated system, and where crop yields are very high. Sale of conserved forage from farms in irrigated and higher rainfed areas of Tunisia to herders in the low rainfall areas there has also proved a profitable venture.

In view of the current water crisis in Central Asia it seems reasonable to assume that much more rangeland will not be preempted for irrigated crop production. Instead, great efforts will likely be made to reduce wasteful water use and to increase efficiency by improving irrigation technology, reorienting cropping systems toward crops with lower water requirements than cotton, establishing realistic user charges for water, and upgrading water infrastructure—especially drainage to alleviate key constraints of salinity and waterlogging. Encouraging progress toward these goals, as well as toward resolving transboundary water issues, has been made since the governments of the region established the ICWC in 1992 and the Aral Sea Basin Program (ASBC) in 1994. However, these groups have focused primarily on irrigation system and river basin issues rather than on water for the rangelands. Problems which merit examination in a regional framework (perhaps along the lines of the ICWC) include

- the regulation and efficient distribution of water points in rangeland systems and the optimum balance between winter and summer grazings;
- arrangements for maintenance of existing and new wells, boreholes, and pumps;
- the scope and potential for water harvesting and storage systems;
- the nature and use of grazing and water user fees by governments or local authorities;
- the establishment of mechanisms for arbitration in the case of disputes over resource use, rights to rangeland, or water resources, and quality control over those resources to prevent water pollution or range degradation; and
- the potential for linkages between irrigated land and range-based livestock to graze stubbles and by-products on irrigated farms, perhaps on a rental basis, a common practice in some WANA countries.

Reaching a consensus on these complex issues will require the involvement of all stakeholders, including government agencies, relevant industries, NGOs, and the private sector. An up-to-date resource inventory will also be essential as a basis for monitoring range quality and use, assessing water requirements for livestock, and developing a water policy for the rangelands. That water policy should include the creation of wells, cisterns, and tanks to protect the resource from degradation, promote efficient movement of flocks, and limit losses from drought.

Land Tenure Issues and Options

In the space of about three generations rangeland use and control in Central Asia has undergone massive changes. What began as a system of traditional communal grazing, based on mobility and open borders, changed to a centrally planned and state controlled system, with collectives and state farms geared to meeting socialist objectives through increased numbers of animals under more intensive management. Then, from the early 1990s, the system changed again with the establishment of five independent republics and a transition toward a market-oriented economy. The rapidity and comprehensive nature of this most recent transformation presents a major challenge to governments and agricultural communities with respect to management, control, and rights to land—especially in the rangelands where aridity and water scarcity limit the scope for settled systems of animal husbandry. Pressure on these lands has increased as a result of the transfer of resources to irrigation projects and mechanized rainfed crop production. Several questions arise with respect to the future of rangelands:

- What has happened to tribal or other prior rights to range use? Do records of those rights still exist as a basis for future policy?

- Is the migratory, communal open-range grazing system still active and viable, or is privatization a better or a feasible option for the future? What should be the criteria for reallocation of lands? Who gains and who loses?
- Should there be an intercountry agreement on range use and management, with appropriate monitoring and perhaps licensing of flock movements and numbers, and administrative support?
- Can answers be found to some of these questions by referring to situations in other geographic/ecological regions?

Probably the most relevant situation ecologically and historically is WANA, where for centuries tribal rangelands were generally collective property, without titles or charts of ownership, based on an understanding among users. Since the 19th century, when states laid claim to the land in most countries, the instruments have varied: individual, collective, state, and common property, including state claims to rights over tribal rangelands.

Since 1994, IFPRI and ICARDA have been working closely with national teams from eight WANA countries to study the nature and impact of property rights in those countries on economic growth, poverty alleviation, and sustainable resource management.[3] This work was based on cross-regional analyses of national situations and on in-depth research of selected rural communities in Jordan, Morocco, Syria, and Tunisia. The results were presented and discussed at an international conference on policies for the LRAs of WANA held in Jordan in September 1997 (Ngaido et al. 1997). The study made the following conclusions:

- The different policy reforms and interventions in those countries focus on both pasture and cropland property rights institutions. With regard to pasture, the concern is to create property rights institutions that foster sustainable management of common resources and induce greater stewardship by rural communities. With regard to cropland, the concern is creation of a property rights environment that enhances farmers' options for greater efficiency in the use of their productive resources and income generation. As such, land policies have efficiency, equity, and sustainability implications for resource access and use.
- Land rights have direct effects on the ability of farmers and communities to improve their resources, and they also influence the income generation options of rural households.

In all countries a positive relationship was observed between land rights and improvements, both in the case of complete (*mulk taam*) or incomplete (*mulk naqqis*) ownership rights. The propensity to improve, however, varies from coun-

---

[3]This work has been supported by funds from the International Fund for Agricultural Development (IFAD) and the Arab Fund for Economic and Social Development (AFESD).

try to country, reflecting the policy environment under which farmers make their productive decisions.

For example, in Jordan, where the Meeri rights are registered and secure, and in Morocco where tribal members hold perpetual use rights to tribal collective land, rights holders can invest in their lands and reap the full benefits of their production. The importance of tenure security is confirmed by the higher propensity to invest in these lands. Moreover, the Moroccan farmers, with perpetual use rights, are now acting as de facto private owners of these lands and are developing or selling them. In some Moroccan communities, farmers have informally divided all the land among tribal owners. However, these developments are said to be a source of dispute and a concern to the government.

In Tunisia, where the government is actively supporting land privatization, tribal rights have evolved quickly to private property, and presently most lands are registered or titled. In Syria, farmers are investing more in their leased state lands and inherited-divided agrarian reform lands, suggesting they expect to secure ownership rights over these lands in the long term. In general, rented or share-cropped lands are the most disadvantaged because the tenants cannot normally use their holding as collateral for credit and therefore have little incentive to invest in long-term improvements to the property.

Thus the ongoing debate on the completeness or incompleteness of land rights in WANA is really about the different options that are allowable under each tenure regime. The best way to address some of the efficiency and equity issues may be to develop land markets that facilitate transfer of resources from inefficient to more efficient producers. Furthermore, this would encourage better tenure arrangements, enabling tenant farmers to invest in more efficient and sustainable practices.

Because of the differences among countries, the participants in the conference developed a simplifying assumption: "any institutional form is of positive value if it provides tenure security to land users long enough for them to make economically rewarding productive investments, yet at the same time avoids degrading the environment and reducing equity." They also produced a table of positive and negative effects arising from secure property rights or the lack there of (Table 6.3). The participants sounded these cautionary notes:

- Property rights and security of tenure do not necessarily guarantee equity or even efficiency, citing state farms and collectives in Algeria and Eastern Europe as examples.
- Land nationalization in WANA has generally had a poor record because governments have failed to install effective systems of rangeland management to replace traditional tribal or communal rights.
- Privatization, especially of individual holdings, could lead to over exploitation of the resource base, generating undesirable externalities. In addition, it could create inequities if the individual holding ceiling is too large, or con-

TABLE 6.3 The positive and negative impacts of property rights or their absence

Secure rights/Positive impacts	Insecure rights/Negative impacts
Provide stability and incentives to make productive and sustainable long-term investments in resource management	Weaken government authority, impede long-term planning and decisionmaking by public and private sectors
Stimulate adoption of improved technology	Limit the scope of technical improvement and increase risks
Improve access to credit by increasing collateral	Reduce access to credit because of lack of collateral
Facilitate individual and community decisionmaking and conflict resolution	Generate conflicts and trespasses over land and water resources
Create or expand land markets and transactions	Increase land fragmentation
Facilitate legal privatization where desirable	Encourage open access to land and land-grabbing to establish rights
Limit incursion of mechanized cultivation onto rangelands and promote better resource stewardship	Facilitate misuse of resources, the incursion of mechanized barley onto rangelands, and overgrazing of the reduced area of remaining rangelands, leading to soil and biomass degradation and eventually to desertification
Encourage better breeding and management of sheep flocks, leading to improved productivity and higher returns	

demn a family to poverty if it is too small.[4] Privatization should not result in fragmentation.

However, the conference participants emphasized the need to consider a broader perspective of privatization than the Western concept of individual holdings. Privatization could also apply to collective, communal, or cooperative systems, and to different village, tribal, or administrative district levels. They con-

[4]For farm privatization in Tunisia, the state subdivided land into units from 100 hectares to 1,000 hectares in size, on the basis of production systems, landscape, and water availability. These were leased by tender for long terms, mainly to groups of owners who had to submit production plans and evidence of their competence before being granted the land. In Central Australia large private ranches are divided into paddocks by their owners to provide better grazing control and to capitalize on local variability in soil and terrain.

cluded that the greatest environmental and social urgency was to establish clearer and more secure rights for the rangelands and identified the following needs:

- Information, education, and appropriate technological packages to ensure sound land use and effective policies
- Efficient management institutions for collective rights integration and recognition of traditional rights and regulations
- Creation of mechanisms and principals for the transfer of government lands to communities and/or private ownership
- The involvement of local leadership and empowerment in decisionmaking
- Establishment of a communication flow between communities and decisionmakers to provide them with essential information

Drought Management and Its Wider Implications for Policy

A large proportion of the land in most countries of WANA and Central Asia lies in low rainfall areas with annual precipitation below 200 millimeters. Land defined as pasture represents more than 50 percent of total area in all five Central Asia countries, and much of the remaining land is too dry or too mountainous to support agriculture. Although there are upland pastures in those countries (with somewhat higher rainfall), among the three states with the largest total areas of grazing, between 70 and 90 percent of their land lies in the arid zones (Table 6.1). This is also true of most WANA countries, Turkey being somewhat of an exception because of its large area of cropland, smaller share of really arid land, and better integration of crops and ruminant livestock in farming systems.

Low rainfall in itself is not necessarily an insuperable problem. As long as precipitation is well distributed and not highly variable within and between seasons, farmers can adjust their cropping or grazing systems and management techniques to deal with it. However, where precipitation is both limited and erratically distributed, with frequent unpredictable and sometimes prolonged periods of drought imposed on aridity, farmers and herders often have difficulty in coping, crop yields can fall dramatically, serious losses of livestock can occur, and farm incomes may collapse. For example, in the devastating 1994/95 drought in Morocco, cereal production fell from the previous year's 9.6 million metric tons to 1.7 million metric tons, disrupting the national economy and forcing the government to seek a loan from the International Monetary Fund. In the prolonged droughts of the early 1980s, 25 percent of the cattle and 30 percent of the sheep in Morocco perished or were sold prematurely; the sheep population in Syria fell by 25 percent (2.5 million head); and, in Iraq both crops and range vegetation failed almost completely. In the drought of 1997, a third of the sheep in Jordan were slaughtered prematurely. And in the current drought in Syria, it is reported that farmers with boreholes are selling water to the local population because of domestic water shortages; the well owners consider this to be more lucrative than using the water to grow crops.

It would be illuminating to compare these examples with developments in Central Asia, to find out whether similar emergencies have occurred there—and if not, why not—as well as to examine reasons for the significant changes in the livestock situation since 1990. Why, for example, have cattle and sheep numbers fallen significantly in Kazakhstan, the Kyrgyz Republic, and Tajikistan (and also in Turkey) but risen in Turkmenistan and remained more or less constant in Uzbekistan? If these changes were not largely climate related, what were the causes, and what do they imply for future policies toward rangelands and livestock in Central Asia?

Some have suggested that, rather than drought, a decline in rural services, livestock owners' lack of cash to buy feed and forage supplies, rising costs of fuel and of trucking animals to grazings, and market failures were at the root of these problems. But the effects of a severe drought would severely compound them.

In this light it would be prudent for Central Asian governments to develop a strategy for the range livestock sector which would (1) provide a longer term approach to drought management than the stopgap efforts of the past; and (2) address other key problems adversely affecting the ruminant livestock economy in the region.

The question here is how to raise the productivity of livestock without inflating numbers to a level that would further degrade the resource base and increase the risks of heavy losses of animals through drought or other emergencies. A second question is how to do so without incurring costly dependencies on feed supplements of the type prevalent in WANA (Oram 1997). Drought should be regarded as a structural feature of the regional climate and an integral component of the production function, not merely as a random emergency; thus its management should be a central element of rural development strategy together with the improvement of rural incomes and poverty alleviation (El Mourid and Moussaoui. 1997).

Experience in other regions where drought is endemic (WANA, South Asia, Australia, and the western United States) suggests that such a strategy for Central Asia should include the following steps:

- Establish a permanent institutional base to manage drought, identify priorities, coordinate actions, improve early warning systems, and supervise drought monitoring. Reliable and timely information on drought conditions and related effects is essential. India's experience with its Drought Advisory Committee (DAC) is relevant (Walker, 1991).
- Develop a national rangeland database supported by a monitoring, evaluation, and information system using modern technology (resource inventory, geographic information system, and early warning capability).
- Establish a strong and effective research network that addresses the major issues of LRAs, so as to increase and sustain productivity and alleviate poverty while reducing damage to the environment. Such a network would help develop baseline information systems on physical, biological, sociological, and economic environments and improve understanding of farmers' and herders' techniques of coping with drought and other emergencies

affecting their resources and their livestock.

- Increase public investment in the development of LRAs, for example, investing in rural public works (especially roads and water supply); implementing water harvesting and land reclamation techniques at farm level; and building schools, health centers, and public services. These measures create employment and can decrease rural out-migration while improving revenues from off-farm activities.
- Encourage the private sector to invest in rural agricultural development of LRAs, for example, by developing small businesses and ecological tourism; mechanizing well-digging; and improving rural transportation for fattening live animals and for marketing feed and livestock. These activities can also increase employment opportunities and improve the livelihood of the local population.
- Implement policies and technologies, such as minimum tillage, to protect the environment, discourage cultivation of marginal lands, increase shrub fodder/agroforestry plantations as drought reserves (alone or in association with crops in alley-cropping systems), apply phosphates to pastures in less marginal areas, conserve biodiversity through improvement of grazing management, and establish a core of protected areas in sites of special importance.
- Develop loan and insurance systems to help farmers incorporate appropriate levels of risk management in individual farms.
- Promote closer crop-livestock integration. This is an important priority both on individual farms and across ecozones (e.g., rangelands to farms in higher rainfed and irrigated zones), but its success depends on integrating several complementary factors. These include establishing secure property rights to individuals or communities and discouraging land fragmentation; improving marketing systems for livestock, inputs, and outputs, with support to producer cooperatives; and promoting fodder crop production and feed storage at the farm level.[5]
- Integrate the provision of veterinary services with the national crop/range/livestock plan, especially with respect to cost-recovery and the control of parasites and major diseases such as sheep pox. Healthy animals are less likely to succumb to drought or freezing temperatures.[6]

[5]Recent developments in feed technology (including urea treatment of straw to improve its palatability and digestibility, and the production of low-cost high-energy feed blocks from crop by-products as a dietary supplement) offer opportunities for improving economic efficiency in the ruminant livestock sector as well as helping to mitigate drought risks.

[6]A controversial issue being debated in WANA is whether to increase sheep fertility in LRAs to achieve higher rates of twinning through hormone treatment. Although economically attractive, this has the obvious risk of animal losses in the more drought-prone areas and could increase degradation from increased grazing pressures.

- Enable herders to destock or restock grazings to cope with climatic or economic emergencies. Such measures are now seen as a key element of sustainable livestock and natural resource management in LRAs, but they depend heavily on access to markets, effective transport, slaughter and cold storage facilities, and an equitable price structure for sale and purchase of animals.

A related issue of particular importance to Central Asia is whether the main market focus for sheep should be for fine wool (in which case the incentive to produce sheep meat may be low); or for meat (Schillhorn van Veen. 1995). It may be worthwhile for governments and for private entrepreneurs in Central Asia to study the market for meat in the WANA countries, given their large and increasing supply deficits and relatively high per capita incomes, as an opportunity for increasing exports from this region.

References

El Mourid, M., and M. Moussaoui. 1997. The experience with drought management policies in WANA. In *Agricultural growth, sustainable resource management, and poverty alleviation in the low rainfall areas of West Asia and North Africa*, eds. Chaherli, N., P. Hazell, T. Ngaido, T. Nordblom, and P. Oram. Feldafing, Germany: Food and Agriculture Development Centre, German Foundation for International Development.

FAO (Food and Agricultural Organization of the United Nations). 1996. *Production Yearbook and Agricultural Statistics*, Vol. 50. Rome.

______. 1997. *Production Yearbook and Agricultural Statistics*, Vol. 51. Rome.

Government of Kyrgyzstan. 1995. *National Environmental Action Plan*. Report No: 13990-KG. Bishkek, Kyrgyz Republic: Government of the Kyrgyz Republic and the World Bank.

Masri, A. 1991. The tradition of hema as a land tenure institution. In Arid land management in Syria. Rome: Food and Agriculture Organization of the United Nations.

Ngaido, T., M. Abbassi, A. Baalbaki, M. S. Bachta, M. Boughlala, P. Hazell, A. Khnefis, M. Mustapha, F. Nassif, C. Nasser, K. Nesheiwat, T. Nordblom, and N. Redjel. 1997. Land improvements and sources of income in the low rainfall areas of the Mashreq and Maghreb regions: Do property rights matter? In *Agricultural growth, sustainable natural resource management, and poverty alleviation in the low rainfall areas of West Asia and North Africa*, eds. Chaherli, N., P. Hazell, T. Ngaido, T. Nordblom, and P. Oram. Feldafing, Germany: Food and Agriculture Development Centre, German Foundation for International Development.

Nordblom, T., and F. Shomo. 1995. Food and feed prospects to 2020 in the West Asia/North Africa regions. Paper presented at the workshop on "Ecoregions of the Developing World," International Food Policy Research Institute, Washington, D.C.

Oram, P. 1997. The influence of government policies on livestock and the environment in West Asia and North Africa. *Proceedings of the International Conference on Livestock and the Environment.* Ede, Wageningen, Netherlands.

Oram, P., and A. Belaid. 1990. *Legumes in farming systems.* Aleppo, Syria: International Center for Agricultural Research in the Dry Areas (ICARDA).

Schillhorn van Veen, T. 1995. *The Kyrgyz sheep herders at a crossroads.* Pastoral Development Network Paper 38. London: Overseas Development Institute.

Taimeh, A. 1997. Environmental threats to the low rainfall areas and their causes. In *Agricultural growth, sustainable natural resource management, and poverty alleviation in the low rainfall areas of West Asia and North Africa*, eds. Chaherli, N., P. Hazell, T. Ngaido, T. Nordblom, and P. Oram. Feldafing, Germany: Food and Agriculture Development Centre, German Foundation for International Development.

Walker, T. S. 1991. Improved technology in relation to other options for dealing with climatic risk to crop production in the semi-arid tropics. In *Climatic risk in crop production: Models and management for the semi-arid tropics and subtropics*, eds. R. Muchow and J. Bellamy. Brisbane, Australia: CAB International.

World Resources Institute. 1992. *World Resources Report.* New York: Basic Books.

7 Trends in Feed and Livestock Production during the Transition Period in Three Central Asian Countries

MEKHLIS SULEIMENOV

Before obtaining independence in 1991, the Central Asian countries were part of the Soviet Union and shared one political and economic system. Agricultural production was organized mainly in the large collectives and state enterprises which, excepting some legal differences, differed very little. For both, decisions on crop-sown area, livestock numbers, and targeted agricultural production for each farm were made by the leadership of the Communist Party of the Soviet Union, the republic, the region, and the district. In turn, the government was obliged to supply necessary inputs and to procure agricultural produce.

Despite their many commonalities, the countries of Central Asia, for various natural, historical, and demographic reasons, have had sometimes significant differences in their agricultural development, even in the Soviet period. This chapter attempts to analyze trends in livestock development during the transition to a market-oriented economy in three Central Asian republics, namely Kazakhstan, Kyrgyz Republic, and Uzbekistan.

Before 1991, agricultural production in Kazakhstan was conducted primarily by very large state-owned enterprises established in the steppe area, with semi-arid climates very much like the Canadian prairies of Saskatchewan, Alberta, and Manitoba. These enterprises were established to produce high-quality wheat. In addition, all grain farms had large livestock operations, and the fodder to feed the livestock was produced largely on the farms. Each farm also had outreach rangeland sites used for the seasonal grazing of livestock and sometimes for cutting hay for winter conservation and feeding to animals. In emergency cases and depending on the scale of the disaster, the Communist Party leadership would transfer fodder (hay, silage, straw) from one farm to another, one district to another, or one region to another. Similarly, during natural disasters (for example, heavy snowfalls) the Central Committee of the Party could summon airplanes, helicopters, trucks, and labor from industrial enterprises and military units. These factors are important to keep in mind while discussing reasons for the demise of the livestock subsector of Kazakhstan.

The Kyrgyz Republic conducts agriculture on a smaller scale. In this mountainous country, the main subsector during the Soviet era was sheep production on highland rangelands. Other livestock and crop production were of minor significance. Uzbekistan used to be the main producer of cotton on irrigated lands and also intensively grew fruits and vegetables for other parts of the Soviet

Union. Livestock was mostly sheep, primarily Karakul sheep raised in deserts and semi-deserts.

After the breakup of the Soviet Union, the republics of Central Asia started to restructure their agriculture, targeting two major goals: achieving food security and adjusting to market economy requirements. Although the countries encountered many common challenges, their agricultural policies differed significantly for a number of reasons. It is therefore important to see the results of these policies and to learn lessons from them.

Characteristics of Reforms in Agriculture

Kazakhstan

As a result of reforms conducted in 1994-97, by the beginning of 1997, almost 98 percent of agricultural entities in Kazakhstan were privately owned. In the Commonwealth of Independent States (CIS) it is commonly said that there are "privatized" farms and there are "private" farms; that is, there are legally private farms and truly private farms. This chapter divides farms into three major groups: agricultural enterprises comprising various types of collectively run farms, essentially those that were formerly state-owned farms; individual farms privately owned by one person; and subsistence household plots or household operations.

As a result of the reforms, most farmland remains in large agricultural enterprises, with a considerable share of it abandoned as a weedy fallow. Individual farms comprise only about 15 percent of cropland, and individual households have almost no land for farming. Ironically, the large enterprises, which occupy nearly 85 percent of cropland, produced 43.8 percent of agricultural output, while individual farms, with 15 percent of cropland, produced 10.2 percent of output and household plots, with almost no land, produced 46.0 percent of agricultural output in 1997.

Kyrgyz Republic

The Kyrgyz Republic pursued essentially the same concept of reforms, but because of different scales and the type of farming, the results were quite efficient. In particular, before reform, this country did not have large state farms growing rainfed grain. About 78 percent of the cropland is irrigated, and farm units were relatively small. Therefore farms were more easily divided into individually run farms with an average size of 10 hectares of cropland. As a result, the proportion of individual farms is much higher compared with Kazakhstan (Figure 7.1).

Uzbekistan

Uzbekistan pursued a rather different set of reforms, although some results look the same. As distinct from its two neighbors, Uzbekistan did not complete the full

FIGURE 7.1 Distribution of crop land area among private farm types, 1997–98

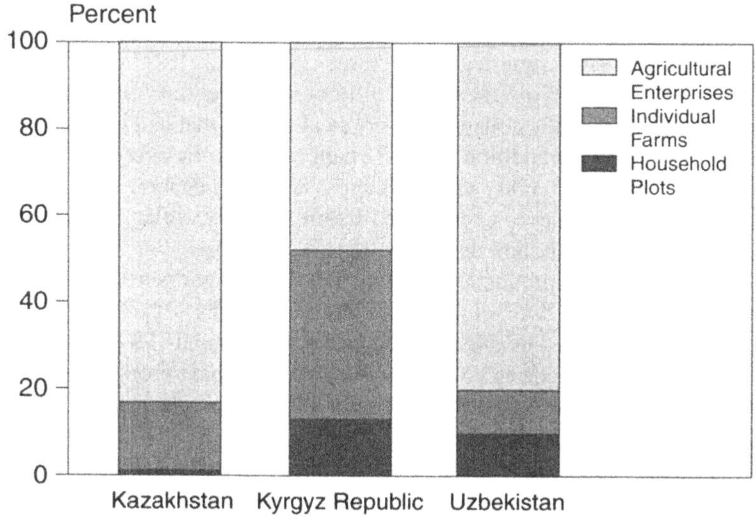

Source: Author's calculations

liberalization of output prices for strategic crops (cotton and wheat) or provide subsidies to cotton and wheat growers. As a result, crop production, excepting fruits and vegetables, remained under government control.

Agricultural enterprises in Uzbekistan are fundamentally different from the other two countries. In Kazakhstan the agricultural enterprises are large, formerly state-owned grain farms run by various kinds of shareholding companies. Former farm laborers became shareholders, working for one person as they had worked for the state before. In Uzbekistan the large enterprise is also a shareholding company, but the shareholders work on the basis of individual family contracts with the farm management. The contracts include an agreement on sharing the inputs and outputs. The incentive for the producer (i.e., the shareholder) is that any crop yield above the agreed-upon, readily achievable level belongs to him. This is not a new way of organizing crop and livestock production, as it was widespread at the end of the Soviet era. In Kazakhstan, for example, "group leasing" was practiced on the basis of a contract with a farm manager. In 1990, farm laborers rented 85 percent of the cropland devoted to cattle farming, 83 percent of that for sheep, 88 percent of that for goats, and 70 percent of that for pigs (Government of Kazakhstan 1991). Some over-reporting probably took place, but clearly leasing was quite widespread in Kazakhstan, as it was in Uzbekistan.

In both Kazakhstan and Uzbekistan, the proportion of individual farms is rather low. In Uzbekistan farm laborers may feel more comfortable working on larger farms. In Kazakhstan, this is not the case, but it is perhaps the lesser of two evils. Individual farming is also quite challenging in Kyrgyz Republic but not as difficult as in Kazakhstan because of milder climatic conditions and the extent of irrigation.

Livestock Production Changes during the Transition Period

Agricultural enterprises and their members have suffered considerably during the transition period. Any significant structural change can lead to difficulties in adjusting to the new environment. The difficulties in Central Asia were extreme as they coincided with the collapse of a political system that had been in place for more than 70 years. In addition, radical changes had to be accomplished in the context of political instability and dramatic changes in economics and finance, which the region had never experienced before, namely, inflation, high interest rates, no access to credit, and unstable prices.

Data on livestock production indicate that the changes were less dramatic in Uzbekistan than in the two neighboring countries (Figure 7.2). Meat production in Kazakhstan decreased by 54.0 percent between 1990 and 1997, and in Kyrgyz Republic, by 26.8 percent; in contrast, in Uzbekistan, meat production increased by 46.9 percent. This same trend was observed in milk production but with much smaller differences between countries.

Wool production was in a much worse shape than meat and milk production in all three countries because of very low wool prices and the absence of established marketing channels. Again Uzbekistan fared better, although a contraction

FIGURE 7.2 Livestock production changes during the transition period in 3 Central Asian countries

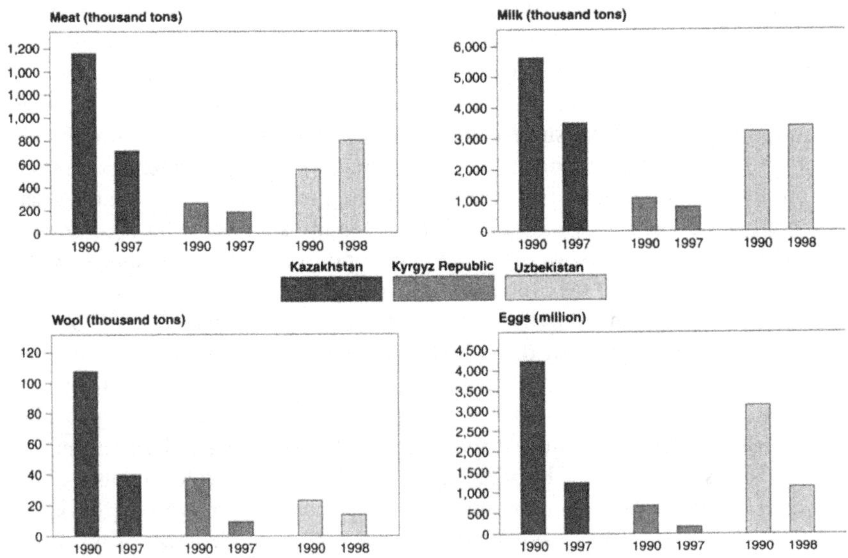

Source: Author's calculations

of wool production still occurred. Egg production suffered even more than wool in all three countries, primarily because, in the past, chicken factories were supplied with imported feed concentrates. This practice stopped immediately after the collapse of the Soviet Union.

The first conclusion apparent from these observations is that the more conservative macroeconomic and agricultural policy of the Government of Uzbekistan resulted in less dramatic losses in agricultural production over the transition period. Another reason for the different performances among the agricultural sectors of the three republics is the harsher climate prevalent in most of Kazakhstan. Winters there last for seven long months, with continuous snow cover and very low temperatures, during which livestock must be kept in heated barns. However, cutbacks in electricity supply and breakdowns in the heating systems have resulted in very high mortality rates.

Yet another explanation for Uzbekistan's better fortune is that its livestock operations were not as large as those in Kazakhstan. Most of its cattle (56 percent) were already kept on household plots, producing 59 percent of milk in 1990 (Rzaev 1990). In the same year Kazakhstan produced 46 percent of its milk on household plots, and the Kyrgyz Republic, 52 percent. This occurred long before the reforms, when household plots were very much involved in livestock production (Table 7.1). In the course of the reforms, livestock production shifted further to household plots in all three countries, not so much because households plots expanded, but more because the number of livestock on large farms contracted (Table 7.2).

In meat and milk production, the share of household plots is considerably higher in Uzbekistan; large agricultural enterprises still play a role in Kazakhstan, and individual farms do so in the Kyrgyz Republic. The Uzbekistan case is explained by the fact that the livestock subsector receives no subsidies from the government. As a result, the proportion of individual farms in livestock produc-

TABLE 7.1 Share of household plots in agricultural production in three Central Asian countries, 1990

	Kazakhstan	Kyrgyzstan	Uzbekistan
Grain	0	4.5	8.6
Potatoes	55.9	51.8	47.8
Meat	33.0	44.9	36.0
Milk	46.0	51.5	59.1
Eggs	92.0	n.a.	37.2

SOURCE: Author's compilation from unpublished official records.

NOTE: n.a. indicates not available.

TABLE 7.2 Distribution of livestock production on various farm types in three Central Asian countries, 1997

	Kazakhstan	Kyrgyz Republic	Uzbekistan
Farm type		Meat	
		(percent)	
Agricultural enterprise	27.0	3.6	8.9
Individual farm	5.5	20.1	1.3
Household plot	67.5	76.3	89.8
		Milk	
Agricultural enterprise	12.2	6.7	6.8
Individual farm	4.8	21.6	1.5
Household plot	83.0	71.7	91.8
		Eggs	
Agricultural enterprise	53.0	1.2	34.9
Individual farm	1.0	13.4	0.6
Household plot	46.2	85.4	64.5
		Wool	
Agricultural enterprise	6.1	5.1	27.1
Individual farm	5.8	26.4	0.9
Household plot	68.1	68.5	72.0

SOURCE: Author's compilation from unpublished official records.

tion in this country is insignificant. Conversely, the Government of Kyrgyz Republic does provide some support and so individual farms account for the highest share of livestock production.

In egg production Kazakhstan is distinguished by a rather high share of agricultural enterprise. This demonstrates that some large chicken factories have recovered and produce more eggs than do household plots.

Livestock population changes in all three countries are correlated with production (Figure 7.3). The decrease in small ruminants in Kazakhstan and Kyrgyzstan is striking. Immediately following the onset of reforms, small ruminant inventories began to fall at a rapid rate. In only two years, 1995-96, Kazakhstan lost 14.6 million head or 42.8 percent, including 26.5 percent lost in

1995 alone. This is comparable only to animals lost in Kazakhstan during a period of collectivization in the early 1930s.

In fact, decollectivization damaged the livestock industry of Kazakhstan as severely as did collectivization. First, most of the sheep distributed to farm members were used as barter to buy goods because the farmers had such low earnings from farming. Second, sheep sometimes had to be slaughtered because the farmer could not arrange grazing on outreach rangelands. Even large collective farms stopped moving their flocks to seasonal run-away rangelands because of the lack of resources. Third, the wool market collapsed. And fourth, the situation was aggravated by severe winters and high sheep mortality. Under these circumstances, the managers of the large farms started to get rid of their livestock, beginning with their sheep.

Livestock inventories for the various farm types reveal once again that distribution shifted toward household plots (Table 7.3). Most striking are the shares of cattle on household plots in Uzbekistan (81.9 percent), on agricultural enterprises in Kazakhstan (21.4 percent), and on individual farms in Kyrgyz Republic (29.0 percent).

The shares of sheep on agricultural enterprises are noteworthy in Kazakhstan and Uzbekistan, and on individual farms in Kyrgyz Republic. In Uzbekistan most Karakul sheep have been raised on large collective enterprises

FIGURE 7.3 Livestock population changes during the transition period in three Central Asian republics (thousands)

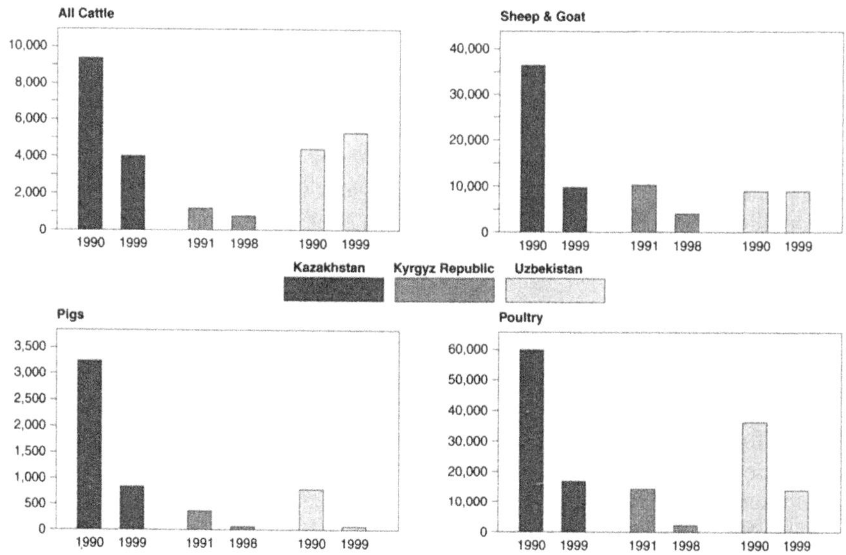

Source: Author's calculations

TABLE 7.3 Livestock inventories on various farm types in three Central Asian countries, 1998

Farm type	Kazakhstan	Kyrgyz Republic	Uzbekistan
		Cattle	
Agricultural enterprise	21.4	9.0	12.9
Individual farm	5.7	29.0	3.3
Household plot	72.9	62.0	81.9
		Sheep and goats	
Agricultural enterprise	26.0	4.8	33.0
Individual farm	8.4	28.3	2.0
Household plot	65.6	66.9	65.0

SOURCE: Author's compilation from unpublished official records.

where the flocks are leased to shepherds, who raise them on contract along with their own flocks. (The same practice is widespread in Turkmenistan.) Usually the ratio of farm stock to the shepherd's own is 50:50.

Livestock productivity indicators on the various farm types show that the milk productivity of cows in the Soviet era was higher on state farms than on household plots, while sheep wool productivity was much higher in private flocks (Table 7.4). The low productivity of privately owned cows in Soviet times is explained by the poor quality of the rangelands around the villages used for grazing and the poor nutritional value of the fodder. These days cow productivity on household plots has improved slightly as a result of better management. Nonetheless, the data indicate that cow productivity remains low. Sheep productivity decreased on both agricultural enterprises and household plots, most noticeably on the latter, which could be explained by poor feeding and breeding.

Fodder Production Trends

The trends in the cropland area allocated for fodder crops are largely the same across the three countries (Table 7.5). The general trend is a reduction in the area sown to fodder crops. Kazakhstan has reduced its sown area of all crops by 37.9 percent, reducing grains by 33 percent and fodder crops by half. At the same time, huge cropland area is left as weedy fallow. Out of 26.6 million hectares of arable land in 1998, only 18.5 million hectares were sown. Of the remaining 8.1 million hectares, 2-3 million hectares were cultivated as summer fallow, and 5-6 million

TABLE 7.4 Livestock productivity on various farm types in three Central Asian countries (various years)

	Kazakhstan		Kyrgyz Republic		Uzbekistan	
Farm type	1990	1997	1992	1995	1990	1998
	Milk (kilograms per cow)					
Agricultural enterprise	3,055	1,557	3,049	1,513	1,950	1,020
Individual farms	n.a.	1,559	2,133	1,701	n.a.	962
Household plots	1,679	1,559	1,778	1,973	1,479	1,630
	Wool (kilograms per sheep)					
Agricultural enterprise	2.6	2.4	3.0	2.4	2.8	1.5
Individual farm	n.a.	3.0	n.a.	3.0	n.a.	0.3
Household plot	5.6	3.0	6.4	3.3	4.9	2.3

SOURCE: Author's compilation from unpublished official records.

NOTE: n.a. indicates not available.

TABLE 7.5 Change of crop-sown area during the transition period in three Central Asian countries

Crop	Kazakhstan	Kyrgyz Republic	Uzbekistan
	(1997 as a percentage of 1990)		
All crops	62.1	92.1	104.2
All grains	67.0	127.2	168.8
All fodder	49.2	47.7	53.1
Perennials Forages	95.8	57.6	60.0
Annual forage	19.0	n.a.	n.a.
Maize (silage)	12.0	31.0	80.5
Cotton	86.7	83.3	81.8

SOURCE: Author's compilation from unpublished official records.

NOTE: n.a. indicates not available.

hectares were left fallow. In addition, about 7 million hectares of marginal lands were used for cropping a few years ago. Now this area is excluded from cropland but not added to rangeland. So, there are 12-13 million hectares of weedy fallow land, which could be put to perennial grasses and used as improved pastures. Currently, these lands belong to agricultural enterprises with no resources to develop them. A special project should be developed for this purpose, providing support to livestock farmers and allocating these lands to them.

Kyrgyz Republic and Uzbekistan have reduced fodder crop-sown area because both increased grain growing area. Remarkably, both countries reduced perennial forages, primarily alfalfa under irrigation, for the sake of grain area. Conversely, Kazakhstan did not touch perennial forages, which are predominantly grasses (Crested Wheat grass), but cut back annual grasses by 80 percent (almost 3 million hectares) and maize for silage by 88 percent (2 million hectares). Now all this is weedy fallow. Kyrgyz Republic and Uzbekistan had almost no annual forages, but these countries reduced maize for silage and green matter by 69 percent and 19.5 percent, respectively.

Such policies cannot be justified. If Kyrgyz Republic was able to return abandoned cropland area to annual grasses, it is possible for the two other countries to look for other sources of animal feed as well. Most probably they will decide to increase fodder production. This can be accomplished partly through better varieties and technologies, but area must increase as well. Kyrgyz Republic has already realized this necessity and is increasing area under alfalfa. In Uzbekistan the trend is to use fodder crops as double cropping after harvest of wheat, but new zero-tillage technologies should be developed for this purpose. Forages are also essential for better crop rotations to improve soil fertility and break the continuous use of the land for grains or cotton (Table 7.6). The share of fodder crops fell considerably in all countries and is especially low in Uzbekistan. Increasing the area under alfalfa will improve soil fertility and provide better quality forage for livestock.

All three countries have vast rangeland areas. Kazakhstan has rangelands in the dry steppe zone, suitable for grazing horses, beef cattle, and sheep; in semi-deserts, suitable for grazing fat-tail sheep, horses, and beef cattle; and in deserts, suitable for grazing Karakul sheep, fat-tail sheep, and camels. There are also some rangelands in foothill and mountainous zones. None of these resources are used properly. In spite of dramatic declines in livestock, overstocking of rangelands still increased because only 30 to 40 percent of rangelands have been used for grazing.

Several factors constrain the use of the remaining rangelands. First is the lack of water points. In the past, special governmental services took care of wells on rangelands. That has stopped because no money has been budgeted for it. Second, most outreach rangeland areas used as seasonal pastures by large state-owned farms are no longer used. Smallholders cannot migrate with only a few sheep; and even for large flocks the long distances are an insurmountable obstacle because of safety problems and the lack of a system of run-away ranges.

TABLE 7.6 Change in cropping composition during the transition period in three Central Asian countries

	Kazakhstan		Kyrgyz Republic		Uzbekistan	
Farm type	1990	1997	1992	1995	1990	1998
			(percent)			
All grains	66.4	71.6	41.5	57.3	24.4	39.5
All fodder	31.4	24.9	49.5	25.7	24.0	12.3
Perennial forages	13.0	20.0	29.5	18.4	17.2	6.7
Annual forages	9.9	3.0	n.a.	n.a.	5.5	3.0
Maize (silage)	6.5	1.2	5.5	1.8	2.2	2.1
Cotton	0.3	0.5	2.3	2.1	46.5	36.5

SOURCE: Author's compilation from unpublished official records.

NOTE: n.a. indicates not available.

Kyrgyz Republic has the same problem with 9 million hectares of highland rangelands, most of which have not been grazed since independence. A sound policy should be developed to encourage rational use of rangelands. This cannot be done without governmental support. While developing a rangeland utilization policy, land tenure should also be addressed. In some developed countries (United States, New Zealand) rangelands are state-owned and leased for long terms under favorable conditions provided a tenant will comply with soil conservation recommendations.

Although the rangelands of Uzbekistan have been used more properly because a state-controlled system is still in place, policies must be developed to prevent soil degradation. The widespread sowing of grain crops on marginal lands contributes to soil erosion.

Recovery of the agricultural sector is impossible without policies providing support to producers. The example of Kazakhstan, a country with large resources, shows that in the course of transition the government did not invest in agriculture (Table 7.7). The share of agriculture as a percentage of gross domestic product fell from 33.9 percent in 1990 to 8.4 percent in 1998. This decline paralleled a drop in the share of major agricultural assets in the total economy: tractors and combines contracted by two-thirds and by half, respectively. The remaining machinery is aging, having been manufactured 10 to 15 years ago.

The livestock industry turned unprofitable right after radical reforms began (Table 7.8). One of the major reasons for this catastrophe was an absence of agricultural policy. The program of reforms consisted primarily of schedules on privatization and liberalization determined with the active participation of the World Bank and International Monetary Fund. These reforms were worse than the policies under leadership of the Communist Party. For example, in Kazakhstan, a pol-

TABLE 7.7 Investments in the farm sector of Kazakhstan

	1990	1993	1994	1995	1996	1997	1998
Share of major assets of agriculture in total economy (percent)	n.a.	n.a.	16.5	14.9	12.9	8.9	n.a.
Share of agriculture in GDP (percent)	33.9	16.4	14.9	12.3	12.2	11.5	8.4
Tractors available (as a percent of 1990)	100.0	96.0	89.0	60.0	48.0	33.0	n.a.
Grain combines (as a percent of 1990)	100.0	89.0	78.0	70.0	60.0	47.0	n.a.
Fertilizer applied (as a percent of 1990)	100.0	34.0	10.8	5.4	4.3	2.5	3.0
GAO[a] (as a percent of 1990)	100.0	93.0	73.0	55.0	53.0	52.0	42.0

SOURCE: Author's compilation from unpublished official records.

NOTE: n.a. indicates not available.

[a]GAO is gross agricultural output.

TABLE 7.8 Profit (loss) margin of agriculture in Kazakhstan, 1996

	Profit (loss) (in millions of tenge)	Profit (loss) margin
Crop	(923)	(1.9)
Grain	658	1.7
Sunflower	130	49.1
Sugar beets	(3.2)	(58.3)
Potatoes	90	6.7
Livestock	(17.915)	(43.7)
Meat	(8.412)	(48.4)
Milk	(4,365)	(43.8)
Wool	(1,269)	(63.2)
Eggs	(73)	(1.6)

SOURCE: Author's compilation from unpublished official records.

icy of taxing the agricultural sector in 1992-94 left the sector largely insolvent. In an effort to survive, farm mangers started to sell assets, including machinery and livestock. While agriculture accumulated debts, input supply became a lucrative business. Insolvent producers could buy inputs only as barter exchange. Under these circumstances, input suppliers established terms of exchange that were unfavorable for farmers and destroyed the farm sector completely.

Discussion

The livestock industry became so unprofitable during the transition that livestock was almost eradicated on commercial farms. This is why animals survived only on household plots. But this type of farming has no future in the market economy. There is no competition among household plots because they are subsistence, not commercial, operations. Although they can market any excess they produce, a one-cow operation will never make progress.

The large agricultural enterprises are remnants of the Soviet system, and they do not fit the market-oriented economy either. Management of these enterprises remains in the hands of one person, and stakeholders act as hired labor, delivering management rights to a manager-president.

In Kazakhstan there are many private, large farms owned by one person. Some of them are quite successful. But, even though farm laborers are doing much better on these farms than on large agricultural enterprises, this type of farming also looks unattractive. This becomes a latifundium with great social inequity wherein one person owns 51 percent of the shares and several thousand hectares of land while several hundred farm laborers own the rest.

The other case in Kazakhstan involves the movement of commercial input companies into the farming business. After becoming rich at the expense of agricultural producers, these companies started to buy elevators, and then management or land use rights to farms. Some companies have already accumulated management and land use rights on hundreds of thousands of hectares. They invest in farming, expecting returns in the future. Again, these companies are buying entire large farms with the old structure in place. Agricultural policy should be reoriented to promote the establishment of individual farmers, renting adequate parcels of land to farm profitably.

Unfortunately, individual farms—the only farm type which remains family-run in developed countries—are not successful in any of the three Central Asian republics reviewed in this chapter. The major problem is the difficulty of establishing such farms. On the one hand, managers of large collective farms do not encourage members of their enterprise to leave and run their own operation because that would conflict with farm management. (Usually, newly established individual farms interact with a main enterprise because they need inputs, services, storage, space, parts, and the like.) On the other hand, while the government legally promotes the establishment of individual farms, it is not able to provide them adequate financial aid. Therefore, most individual farms are hardly surviving. A sound agricultural policy should be developed to help this type of farm be competitive in the market. Incidentally, most individual farms do not deal with livestock because it is unprofitable and real private farms cannot afford to run an unprofitable business.

Under present conditions, individual farmers cannot compete if they are not well established. How can they succeed on farms of inadequate size; with inadequate machinery; without access to credit, input supplies, or marketing cooperatives; and unable to get good farmgate prices? Agricultural policy should provide an environment to ensure a rapid development of this infrastructure.

The former Soviet Union (FSU) is a unique region where countries developed for more than 70 years in a completely different political and economic environment. All people of the FSU were used to following decisions made at the top level, and Communist Party organizations at all levels were responsible for implementation. Now the Communist Party has been removed from the scene, and the government has cut back its involvement in many areas.

References

Government of Kazakhstan. 1991. *Regional statistical yearbook of Kazakhstan.* Almaty: National Statistical Agency of the Republic of Kazakhstan.

Government of Kazakhstan. 1998. *Agriculture, forestry and fishery of Kazakhstan in 1997.* Almaty: National Statistical Agency of the Republic of Kazakhstan.

Government of Kyrgyzstan. 1998. *Kyrgyzstan in figures in 1997.* Bishkek: National Statistical Committee of Kyrgyz Republic.

Minmacroeconomstat. 1998. *Livestock production in 1998.* Statistical bulletin. Tashkent: Republic of Uzbekistan.

Minmacroeconomstat. 1999a. *Sowing areas of agricultural crops in 1998.* Statistical bulletin. Tashkent: Republic of Uzbekistan.

Minmacroeconomstat. 1999b. *Yields of agricultural crops in 1998.* Statistical bulletin. Tashkent: Republic of Uzbekistan.

Rzaev, M. 1990. *Agroindustrial complex of USSR: Challenges and outlook.* Tashkent, Uzbekistan: Ministry of Agriculture.

8 Kazakhstan's Wheat System: Priorities, Constraints and Future Prospects

ERIKA MENG, JIM LONGMIRE,
AND ALTYNBECK MOLDASHEV

As a former republic of the Soviet Union, Kazakhstan played a central regional role in supplying wheat to the rest of the country. Although traditionally not an agricultural society, large areas of steppe land were brought into production during the Virgin Lands campaign carried out between 1954 and 1964. Kazakhstan subsequently became the third largest grain producer and second largest net exporter among the Soviet republics (Pomfret 1995). However, independence in 1991 triggered tremendous changes in Kazakhstan's political and economic circumstances. The country is in the midst of making the transition from a command and control economy toward one that is increasingly market-oriented. In the partially dismantled planned economy of present day, decisions concerning levels of input use and of wheat output—which were previously determined by an economic plan and relayed through government channels to production units—now depend to a much greater extent on other types of signals. These signals may be clearly transmitted in individual components of the wheat system. For example, producers receive information to make decisions on levels of input use, however, a sense of coordination within the system as a whole, and of its integration with the rest of the economy, is not always evident.

Increased wheat productivity and higher export levels are two of the goals most frequently expounded by members of the agricultural research community. However, these objectives have proved difficult to achieve in the years since independence. Area planted to wheat, wheat production levels, wheat yields, and the quantity of wheat available for export have all decreased considerably. In the current production environment, the expressed objectives have not been translated into corresponding incentives at the production level. However, clearly an increase in production levels and productivity must occur before an increase in export levels can take place, if indeed increased exports are a government priority.

The baseline findings through the year 2020 generated by the International Food Policy Research Institute's (IFPRI) International Model for Policy Analysis of Agricultural Commodities and Trade (IMPACT) indicate that the Central Asian region will be an overall net importer of cereals (Pandya-Lorch 1999). The IFPRI simulations take into account the general restructuring that has occurred in the region and overall trends in research expenditures, as well as other factors affecting crop production and consumption (Rosegrant, Agcaoili-Sombilla, and Perez 1995). However, because much of this restructuring is taking the country into uncharted territories, changes and corrections are constantly occurring. The projections have also not yet been broken down by individual country. Kazakhstan is unique in Central Asia in terms of the extent of its involvement in wheat production. This chapter's objective is to provide an overview of Kazakhstan's wheat system and to identify policies and constraints that currently influence the operation and development of the system. In light of present conditions in Kazakhstan, what can we expect from its wheat system and what role can we expect the country to play in future developments of the world wheat market? Is it currently a realistic goal for Kazakhstan to regain its position as a major regional exporter of wheat within the existing policy environment? If not, what kind of policy role on the part of the government will be necessary?

This chapter will not be able to answer all those questions; however, it provides an opportunity to discuss relevant factors affecting production decisions and productivity and to present a framework for initiating the analysis. We examine first the present operating environment of the wheat system in the context of the goals of increased productivity and export levels that have been expressed. We focus particularly on productivity trends and the technological possibilities and constraints influencing wheat supply. Because successful research and extension efforts determine the boundaries of the production-possibility frontiers and set guidelines for the inputs necessary to achieve those frontiers, the prospects for increased wheat production in Kazakhstan are indisputably affected by government agricultural and research policies. Research and available technology, however, only partially determine what actually takes place on farms. Wheat producers will make decisions partly on the basis of their knowledge of the existing scientific possibilities, but they also need to consider budget constraints and trade-offs in time allocations and in the purchase of required inputs (Timmer 1987).

These trade-offs and constraints are often influenced by policies and events outside the wheat system. It is consequently also important to examine relevant factors shaping intrasectoral linkages and overall macroeconomic conditions (and the implicit government priorities implied by them) in determining the operating environment of the wheat system (Timmer 1987). Policies targeted toward the agricultural sector and the economy in general affect the wheat system both directly and indirectly, and to achieve a specific objective, they ideally should be coordinated to some extent in order to avoid conflicting results.

Wheat Demand

Figure 8.1 shows wheat production, exports, and consumption during the years prior to and following independence. An overall downward trend in the last several years is immediately evident in total production and export levels.

In contrast, domestic consumption of wheat for food use in Kazakhstan has remained relatively stable over the last decade. The level of wheat production plays a major role in determining the quantity of wheat available for export.

FIGURE 8.1 Kazakhstan wheat production and use, 1987-99

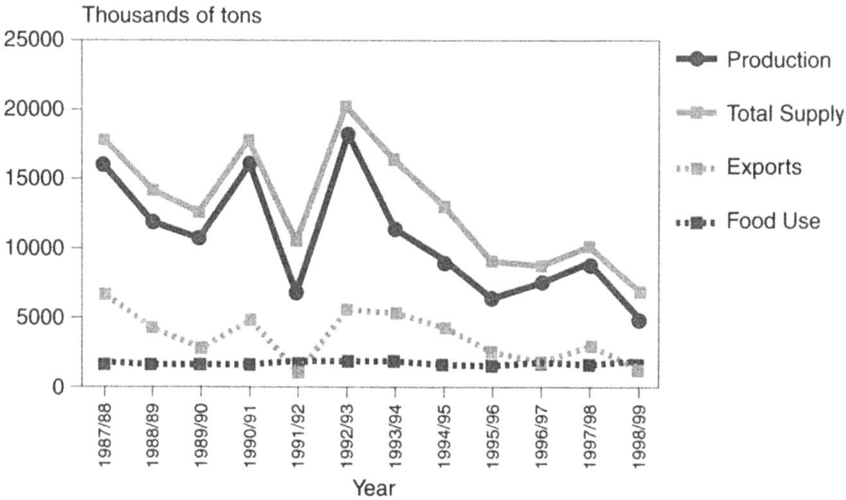

Source: Longmire and Moldashev, 1999

Aggregated IMPACT projections for the former Soviet republics suggest that a combination of slow population growth, slow economic growth, and low income elasticities for cereals will keep the growth in demand for cereals in this region relatively slow (Rosegrant, Agcaoili-Sombilla, and Perez 1995). The population of Central Asia is projected to reach more than 70 million in 2020; Kazakhstan's population in particular is projected to increase approximately 11.3 percent between 1995 and 2020 from 16.8 million to 18.7 million (Pandya-Lorch 1999). Although Kazakhstan is one of only two Central Asian countries expected to experience an increase in the average annual rate of population change between the time periods 1995-2000 and 2015-2020, the rates of change within both these periods nevertheless remain the slowest in the region. IMPACT's projected 3 percent average annual income growth for Central Asia between 1995 and 2020 lies between the projections for non-Central Asian former Soviet republics and for countries in Eastern Europe. All projections for these regions, however, are considerably lower than the growth rates projected for South and Southeast Asia (Pandya-Lorch 1999).

IMPACT findings disaggregating the Central Asian region from other former Soviet republics project that the region will increase its demand for cereals by 33 percent and its demand for meat by 47 percent between 1995 and 2020. However, these amounts account for less than 1 percent of the total global increase in cereal and meat demand for the time period (Pandya-Lorch 1999). They translate into an increase in per capita demand of 10 kilograms for cereals and 5 kilograms for meat and are considerably lower than comparable projections for other regions in the world, particularly Southeast Asia (Pingali and Rosegrant 1998). The increase in demand for cereals in Central Asia can be further decomposed into a 37-percent growth in demand for feed cereals and a 32-percent growth in demand for food cereals. Nevertheless, their relative shares of the total cereal demand (at approximately 59 percent and 21 percent, respectively) do not vary between 1995 and 2020, the two years examined in detail (Pandya-Lorch 1999).

Given the relatively small changes projected for population growth and cereal consumption needs through 2020, fluctuations in productivity and total output will apparently continue to be the primary determinant of export levels. The decade's highest level of exports took place following the bumper crop year of 1992/93; exports subsequently fell to almost 50 percent of that level in 1996/97. They increased slightly the following year only to fall once again during the drought year of 1998/99. Their strong correlation notwithstanding, total output and export levels are the result of the interaction of numerous factors, both inside and outside the wheat system. We now examine institutions and factors within the wheat system that directly affect the supply of wheat.

Wheat Supply

Institutions for Wheat Research and Policy

Two of the principal organizations responsible for wheat-related policies in Kazakhstan are the Ministry of Agriculture and the National Academic Center of Agrarian Research (NACAR) under the Ministry of Science and Higher Education. The Ministry of Agriculture is organized into three main departments: Marketing and Policy, which houses groups for specific crops and livestock; Strategic Planning; and Finance. The Ministry of Agriculture is responsible for policy implementation as well as administrative and technical services and maintains representation in each oblast and rayon (political subdivisions) (World Bank 1999). Responsibilities of the ministry, as well as its staff numbers, appear to have decreased markedly with the shift to a more market-oriented economic system and the corresponding disappearance of centrally planned activities. Its involvement in seed production and distribution has dwindled to a bare minimum (World Bank 1999), as has its role in the procurement of wheat output (Longmire and Moldashev 1999). Fertilizer, herbicides, pesticides, and other inputs recommended for wheat production are no longer subsidized by the government. However, the ministry does participate in a fund to support elite seed producers and will intervene with grain and seed replacement in situations of crop failure, such as the 1998 drought (World Bank 1999). The evaluation of new varieties prior to their release also continues to fall under the ministry's jurisdiction. The ministry does not, however, actively take part in any scientific research, and a formal extension service does not currently exist under its auspices (World Bank 1999).

Primary responsibility for agricultural research and the technological improvement of wheat production takes place in the system of research institutes and experimental stations and farms administered by NACAR. In addition to NACAR's efforts, wheat improvement research is undertaken by the Kazakhstan National Academy of Sciences, also under the jurisdiction of the Ministry of Science, and by Kazakhstan's agricultural universities. NACAR, however, is officially responsible for the overall coordination of agrarian research in the country (Satybaldin et al. 1999). NACAR is also involved in the dissemination of research results; but given the lack of a coordinated extension network (Satybaldin et al. 1999), the transfer of information has been recognized as a major weakness.

Nevertheless, Kazakhstan possesses both the technology and the scientific capacity to develop the technology to attain much higher yields than those presently observed. Prior to independence, a well-trained corps of scientists bred new varieties and formulated detailed agronomic recommendations for wheat production (Satybaldin 1998; Shegebaev 1998) in collaboration with research institutes and experiment stations throughout the Soviet Union. Varieties released in Kazakhstan by local institutes accounted for an estimated 25 percent of the total cultivated area (World Bank 1999).

The research infrastructure from the Soviet era remains in place, now administered by NACAR, although its access to resources has been drastically reduced. The number of agricultural scientists decreased by 50 percent during the 1990s, and in some institutes, personnel levels have shrunk by as much as 75 percent (Longmire and Moldashev 1999). A large number of scientists departed the country of their own accord following independence, and quite possibly some streamlining was overdue. Nevertheless, the large personnel losses and funding cuts during the last decade have placed severe constraints on the breadth and scope of research and extension activities. Funds available in 1998 for research coordinated by NACAR totaled $7.7 million, approximately 0.07 percent of Kazakhstan's gross domestic product (GDP). Budget allocations were reduced another 35 percent in 1999 (Satybaldin et al. 1999).

Furthermore, many of the linkages among institutes established during the Soviet era were broken off after independence. Since then, scientists in Kazakhstan have worked in relative isolation and have been forced to forego potential mutually beneficial opportunities from the exchange of information and materials. The lack of investment for scientific work and the decline in the numbers of scientific personnel have been major factors contributing to the decline of wheat productivity since independence (Longmire and Moldashev 1999). The probability of future advances in wheat productivity will certainly be negatively affected if funding levels remain minimal.

Other Recent Trends in Wheat Productivity

Another major technical factor contributing to the sharp decline in wheat productivity following independence is the accumulated degradation of the resource base from years of unsustainable cropping patterns (Longmire and Moldashev 1999). Some of the land most inappropriate for wheat production has been dropped from production, resulting in a corresponding fall in wheat production area and total output. The decline in output, however, cannot be entirely attributed to the removal of marginal quality land. Satybaldin et al. (1999) estimate that 5.4 million hectares of arable land are currently not sown because of unfavorable economic conditions. Output and cultivated area, as well as average wheat yields in Kazakhstan between 1960 and 1998 are presented in Figures 8.2 and 8.3.

The factors above explain in part some of the recent losses in productivity, and they will continue to affect productivity in the future. However, to better assess prospects for future productivity gains in the wheat system, it is important to identify which, if any, of the technologies promoted during the Soviet era is still appropriate for the current production environment and what new technologies are in the pipeline to replace those that are no longer suitable.

Existing Wheat Technology

What technologies for wheat production are available in Kazakhstan? Is the technology suitable for current production conditions, and, if not, what are the technological constraints for increasing wheat productivity? Wheat production predominantly takes place in two broadly defined agroenvironments, the northern spring wheat environment and the southern winter wheat environment (Longmire and Moldashev 1999). The northern regions consist primarily of dryland steppe

FIGURE 8.2 Kazakhstan wheat area and production, 1960–98

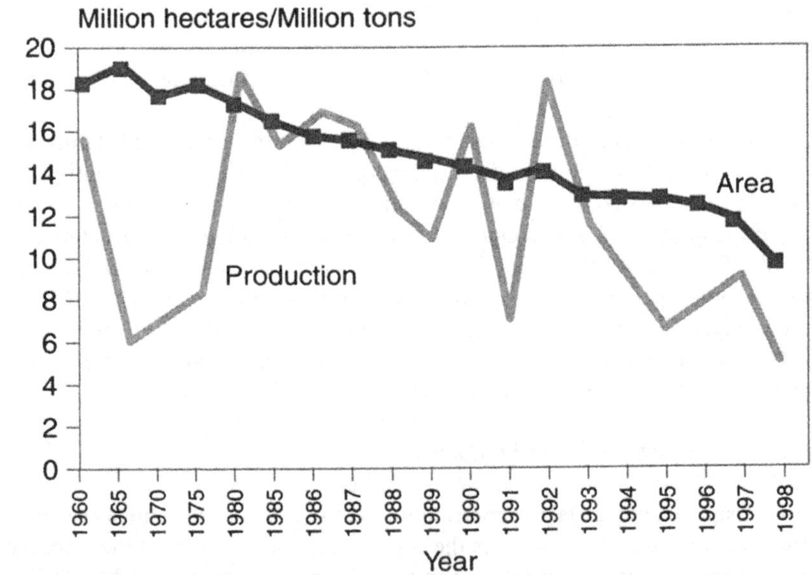

Source: Kazakhstan State Committee for Statistics, FAOSTAT. Cited in Longmire and Moldashev, 1999

FIGURE 8.3 Average wheat yield, Kazakhstan, 1960–98

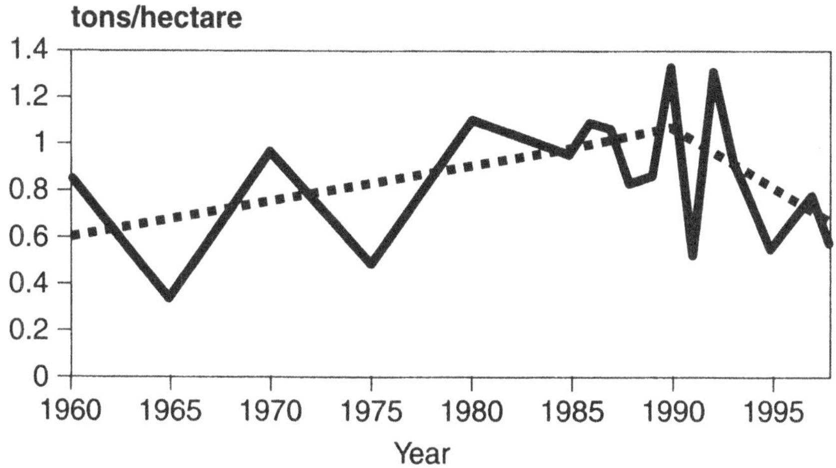

Source: Government of Kazakhstan and FAOSTAT. Data available only at 5-year intervals from 1960 to 1985. Cited in Longmire and Moldashev, 1999.

and are subject to extreme winter conditions and precipitation patterns. In contrast, much of the southern region planted in winter wheat varieties is irrigated. These broad production zones determined the overall direction of research priorities in the past and therefore the nature of the existing technology. The overwhelming majority of wheat cultivated in Kazakhstan is spring type, and more than 75 percent of this total is grown in the northern wheat-producing region of the country (Kaskarbayev 1998). More than 60 percent of Kazakhstan's approximately 25 million hectares of arable land, and consequently much of the arable land in the northern region, is steppe land developed for production activities during the Virgin Lands campaign (Pomfret 1995). Given the significance of the spring wheat region, we focus here primarily on technology developed for that region.

The climatic conditions for wheat cultivation—low and extremely variable rainfall combined with strong winds—required a system of agronomic and soil conservation practices developed specifically to address serious wind erosion problems resulting from cultivation. In accordance with the yield-maximizing objectives of the era, an intensive technological framework suitable for the large-scale production needed to ensure targeted output quotas was developed. The system relied on chisel subsoiling and snow furrowing techniques, harrowing for weed prevention, the timely and intensive use of purchased inputs, and the adoption of the recommended cropping rotation. This technology package contributed to considerably increased wheat yields by enabling moisture retention during the winter and by better utilizing summer rainfall through careful attention to planting time and fertilizer application (Shegebaev 1998). Average yields in spring wheat of more than 1 ton per hectare were achieved in 1990 and 1992 (Figure 8.4).

FIGURE 8.4 Average yield of spring and winter wheat, Kazakhstan, 1960–98

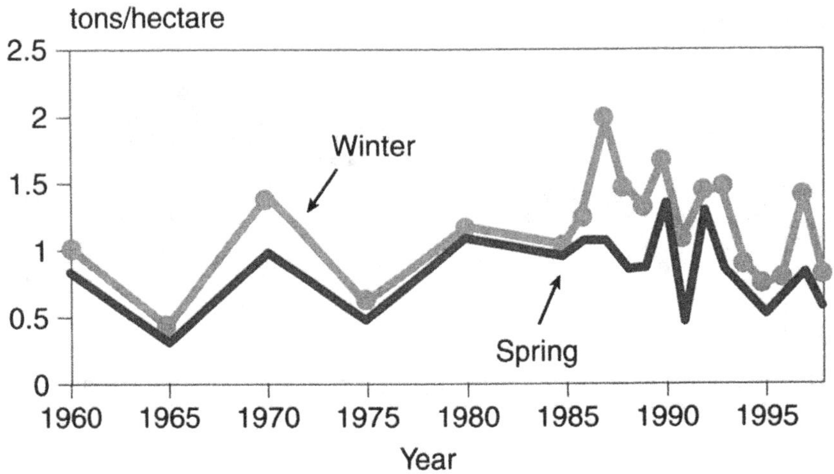

Source: Government of Kazakhstan, State Committee of Statistics. Data available only at 5-year intervals from 1960 to 1985. Cited in Longmire and Moldashev 1999.

During the 1980s input use was intensified again in a concerted effort to increase production levels. Wheat production during the period of the most intense cultivation reached levels ranging from more than 10 million tons to more than 18 million tons (Longmire and Moldashev 1999).

Technology Priorities

Agronomic research. Today there is increased consensus that the intensive level of input use recommended during the Soviet era is at best only locally appropriate. Its widespread use in the past implied a uniformity of growing conditions that is largely unwarranted. Moreover, the objective of maximum yield on land irrespective of long-term environmental consequences can no longer be justified. The intensive use of inputs was one of the primary means used to smooth out variations in growing conditions in the hopes of obtaining desired yields. Scientists continue to face the same natural production constraints that have always existed in Kazakhstan; the inherent challenge of dealing with highly variable, climate-induced cropping conditions, particularly in northern Kazakhstan, remains unchanged. However, research in agronomic methods is increasingly recommended to mitigate the difficult growing conditions. Although more work remains to be done, differences in agroclimatic environments such as soil type and fertility, moisture availability, distribution of weeds, and susceptibility to erosion are now being taken into consideration in new recommendations for agronomic management (Gossen 1998; Shegebaev 1998).

In low-yield areas the role of agronomy in increasing yields relative to that of breeding is particularly important. More efficient management of moisture may be achieved through several different technology possibilities ranging from maintaining crop residue in the field to modifying crop rotation. Evidence suggests that sowing date relative to spring precipitation is also an important factor in yield levels and wheat quality (Sadyk and Morgounov 1999; Shegebaev 1998). Other promising directions for agronomic research include improved crop establishment, improved weed control, efficient fertilizer use, and improved crop rotation (Longmire and Moldashev 1999).

Varietal technology and seed management. The important contribution of new agronomic methods to wheat productivity should not diminish the importance of developing suitable varieties for the region. Breeding objectives for the northern region include the development of varieties that combine high quality, drought resistance, early maturity, and disease resistance. Yield potential, disease resistance, and quality are priority areas identified for improvement in the southern winter wheat region (Movchan 1998). The dry climate, solar radiation, and soils already contribute naturally to the high-quality wheat cultivated in northern Kazakhstan (Movchan 1998), and participation in the high-quality segment of the export market is considered to be a viable possibility for the future (Satybaldin 1998; Shegebaev 1998).

Some research in the development of varieties suitable for different levels of input use intensity has also taken place. The temporary loss of networking capabilities with other research institutes is being addressed; and the establishment of linkages with the International Center for Maize and Wheat Improvement (CIMMYT) and other regional and international research centers introduces a new pool of genetic resources that will help achieve breeding objectives. The exchange of materials and establishment of experimental trials to develop varieties combining new traits with locally adapted ones (for example, day length) has already begun. Considerable scope exists for gaining from collaborative wheat varietal improvement in Kazakhstan (Longmire and Moldashev 1999) and from the renewal of outdated seed testing and seed multiplication equipment and systems (World Bank 1999).

Extension activities. Extension during the Soviet era was handled through the centrally planned research and distribution system. Instructions pertaining to the cultivation of recommended varieties, input use, and planting practices were relayed to agronomists on state and collective farms and from them to farm laborers. Much of the current production structure in Kazakhstan remains unchanged despite changes in name. Nevertheless, the privatization of agricultural land and the appearance of new production units necessitate planning new methods to transfer information. If privatization continues, and particularly if proposed measures to limit farm size are successfully carried out, the development of alternative methods to transmit information effectively from the research level to these new production units will be crucial.

Scientists must also consider the difficult financial circumstances of most farmers both in the design of their research and in their recommendations. Researchers are aware of the changes in production conditions and have begun to address them with new recommendations (Kaskarbayev 1998); nevertheless, some rigidity remains that impedes the modification of the research paradigm carried over from the Soviet era. A result is the danger of irrelevance of some of the work carried out at the research level to the needs at the farm level. Farmers will not adopt recommendations from the research system without the expectation that the cost of the recommendations will be balanced by the benefits of undertaking them. Despite the existence of recommended levels of fertilizer, herbicide, and pesticide use, farms have drastically reduced or even completely eliminated their use of purchased inputs during the last decade. Table 8.1 shows the sharp decline in use of fertilizers, particularly nitrogen and phosphate.

Fertilizer application rates decreased from 13 kilograms per hectare in 1992 to a low of 3 kilograms per hectare in 1995. A similar situation exists with respect to pesticide and herbicide use. The decrease in herbicide use and the number of unused or abandoned fields have contributed to a considerable weed problem that is likely be detrimental to yield levels (Longmire and Moldashev 1999). Given the current economic trade-offs, farmers choose not to follow many of the recommendations regarding input use. There is thus an urgent need for additional investigation into wheat cultivation practices that are simultaneously productive, less damaging to the land, relatively low in input intensity, and cost efficient.

Other Factors Affecting Wheat Production and Productivity

Wheat Prices and Marketing Infrastructure

Barter using grain in exchange for required inputs has become increasingly commonplace. Figure 8.5 shows the level of barter use relative to other wheat marketing means. Longmire and Moldashev (1999) also calculate a weighted average of prices received by farmers through the primary marketing channels for the

TABLE 8.1 Fertilizer use in Kazakhstan, 1992-96

Input	1992	1993	1994	1995	1996
	(thousand metric tons per year)				
Nitrogen	150	86	65	64	63
Phosphate	315	231	50	25	55
Potash	10	7	6	6	6

SOURCE: FAOSTAT 1998 table; cited in World Bank (1999).

FIGURE 8.5 Changing use of wheat marketing channels by farmers in Kazakhstan, 1993–96

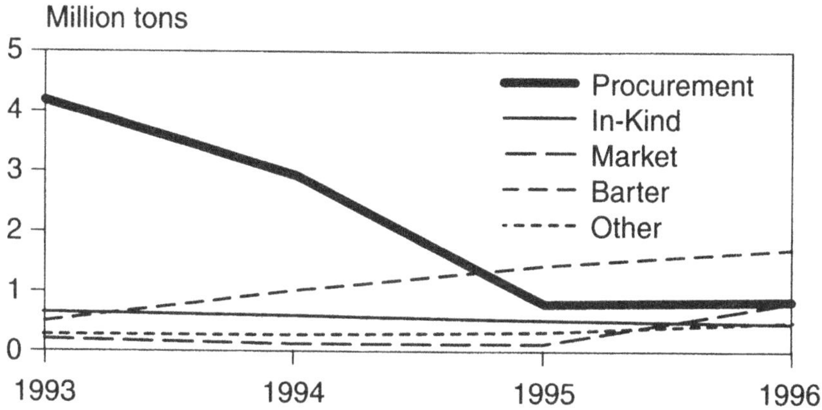

Source: Kazakhstan Scientific Research Institute of Economy and Organisation of Agroindustrial Complex. Cited in Longmire and Moldashev, 1999.

period 1994-98 (Figure 8.6). Recent years have witnessed a decline in real prices, although prospects have begun to improve because of good weather conditions in Kazakhstan and poor harvests in neighboring Central Asian countries (Goletti and Chabot 1999).

Much of the railroad and road infrastructure from the Soviet era is still in reasonable working condition, despite minimal investment in its maintenance and improvement since independence. Kazakhstan also has approximately 11 million tons of elevator and storage capacity (Longmire and Moldashev 1999) which appears to be in relatively worse shape. Nearly one-third of the total production of 1992's bumper output was wasted as a result of poor storage, processing, and distribution facilities (Pomfret 1995). Moreover, this infrastructure was built principally for, and is therefore limited to, the transport of grain and other goods from Kazakhstan to specifically chosen regions of the Soviet Union. Russia and other former republics in Central Asia have continued to be primary export markets for Kazakhstan since independence, and these traditional markets are generally expected to continue to exist. However, many of these countries, particularly Uzbekistan, have intensified their own production of wheat with the help of input subsidies in efforts to achieve self-sufficiency. In contrast to the situation in Kazakhstan, yields have increased in other Central Asian countries (Figure 8.7). Whether or not these yield levels are sustainable remains to be seen. Nevertheless, other export markets will likely need to be targeted, and additional investment in storage and transportation infrastructure will be vital. However, there appears to be little urgency to move in this direction. Plans to build a port for grain transport on the Caspian Sea for easier access to Iran, a potentially large market, have failed to move out of the discussion phase after several years.

FIGURE 8.6 Average wheat price received by farmers, Kazakhstan, 1994–98

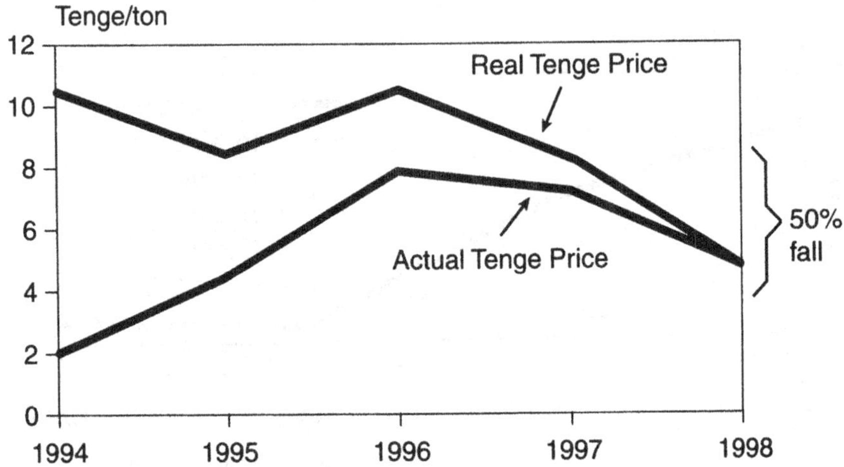

Source: Government of Kazakhstan, State Statistical Office.
Cited in Longmire and Moldashev, 1999.

In recent years, a small number of private grain trading companies have purchased selected components of the country's grain storage system. Some of these companies have also integrated vertically and participate in production activities as well as processing and trade. However, they have little incentive to invest substantially in a marketing infrastructure for export potential that may or may not materialize. Similarly, much of the capacity of the country's fertilizer factories is currently idle because of the lack of demand. Government policies fostering an improved environment for investment in agriculture could generate a response from private firms and perhaps encourage the entry of additional companies. This may be increasingly important because the perception has grown that monopoly power among existing firms has detrimentally affected producer welfare.

Agricultural Sector Linkages

Wheat dominates the agricultural sector with almost 75 percent of the total cultivated area (Satybaldin 1998). Despite the difficulties experienced in the wheat system, some studies show wheat to be more competitive than other available crop alternatives (Satybaldin 1998). Because of its dominance, any changes in the wheat system are more likely to create ripple effects in smaller commodity systems than vice versa. Nevertheless, important linkages to the livestock sector exist, and the sharp decline in the livestock sector could have several ramifications for the wheat system. First, reduced demand for barley and other feed crops could free up arable land, some of which might be used for additional wheat production.

FIGURE 8.7 Average wheat yield, Kazakhstan and other central Asian countries, 1992–98

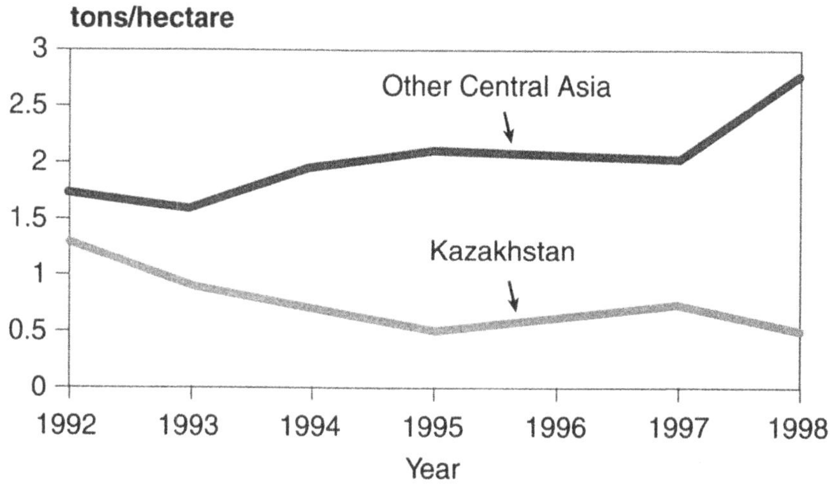

Source: FAOSTAT, cited in Longmire and Moldashev, 1999.

Second, the reduced need to use wheat stubble for grazing purposes could influence the adoption of agronomic techniques to increase moisture retention.

The rural credit system, also a carryover from the Soviet system, has a reputation for favoring collective and ex-state farms (Pomfret 1995). Moreover, no agricultural banking system currently exists in the country. For small farmers in particular, the absence of a method for direct financing is very limiting. However, even large production entities are routinely strapped for credit. Consequently, crucial investment in new machinery and other inputs has not taken place. Because the use of machinery will continue to play a role in Kazakhstan's wheat production area (Satybaldin 1998), the inability to replace or repair existing machinery and inputs is a major productivity constraint.

Wheat and Macroeconomic Policy

The current and future potential of the wheat system also cannot be isolated from the general macroeconomic policy environment. Long-term issues of priorities and funding for agricultural research and education are largely conditioned by government policy toward the agricultural sector relative to other sectors. Agriculture has often been overlooked in favor of government priorities in the oil and gas sector. However, the rural population made up 40 percent of the total population in 1998, and agriculture contributed 13 percent of GDP the same year. Although it commands a small share of total exports (9.3 percent of total exports in 1997) relative to the oil and gas sector (Longmire and Moldashev 1999), agriculture provides the country with the largest source of foreign exchange earnings that is not related to an extractive industry (Figure 8.8). To maintain commodity exports, not only must the producers be competitive in international export markets, but the exporting country must also develop and maintain a reputation among its potential customers for quality and reliability (Timmer 1987). The reputation, as well as the stability itself, is difficult to achieve in the absence of some form of government support.

In determining the distribution of scarce funds among the various sectors of the economy, the effects of budget allocations and priority setting on the attitudes, behavior, and productivity in the agricultural sector cannot be ignored. Agriculture continues to be an important sector of the economy in terms of population, employment, income, and food security. The ramifications of continued unattractive prospects in agriculture, such as increased rural to urban migration rates, are likely to affect many aspects of Kazakh society in a manner potentially inconsistent with overall government objectives for the development of the country.

Competitiveness and Productivity in the Wheat System

In a comparison of production costs for a high-input scenario typical of the Soviet era and a low-input scenario representative of current conditions, Longmire and Moldashev (1999) found the high-input costs per ton to be approximately 20 per-

FIGURE 8.8 Structure of Kazakhstan exports, 1996–97

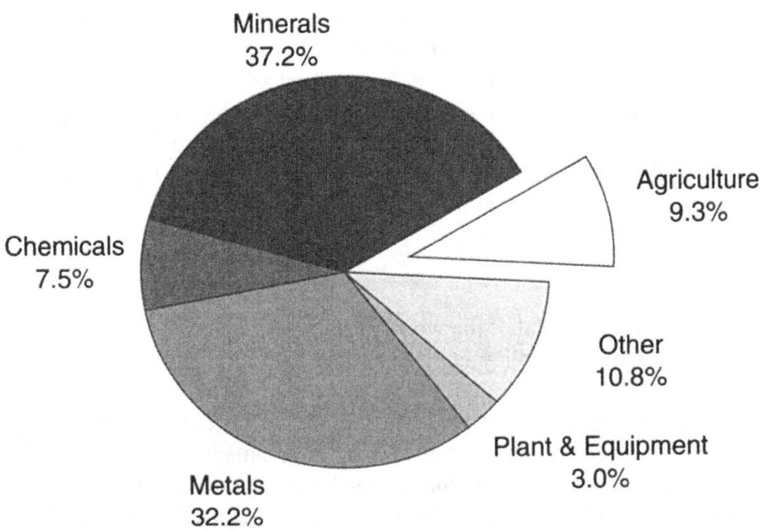

Source: Kazakhstan Economic Trends, First Quarter 1998.
Cited in Longmire and Moldashev, 1999.

cent higher than those incurred using the less intensive input technology, despite the lower yields achieved. Nevertheless, both of the estimated costs per ton, regardless of the technology used, exceed the on-farm parity price of US$100 for 2010 estimated by Longmire and Moldashev. They concluded that wheat will not be competitive with current yields and technology use and would be even less so with the use of an intensive production technology (Longmire and Moldashev 1999). These conclusions reinforce the need for low-cost productivity improvements from the research and extension system in Kazakhstan.

Longmire and Moldashev (1999) also assessed what changes would occur in the competitiveness of wheat production in Kazakhstan as a result of the productivity growth contributed by crop improvement, gains in marketing efficiency, and a decrease in interest rates. Crop improvement would affect costs per ton through yield gains while the two latter scenarios would directly decrease costs of production. With estimated yield gains of 300 kilograms (kg) per hectare from varietal improvement and improved agronomic practices, they concluded that wheat cultivated with relatively low-input technology would be competitive at a parity price of US$80. Wheat production would begin to be competitive at a parity price of US$100 with yield gains of 100 kg per hectare. However, the use of higher input technology requires dramatically higher yield gains, estimated at 700 kg per hectare and 600 kg per hectare respectively, to maintain competitiveness at the parity prices of US$80 and US$100.

Improvements in marketing efficiency through infrastructure development would reduce both costs in acquiring inputs and in marketing output. Longmire and Moldashev (1999) estimated the savings in production costs per ton to be equivalent to yield improvements of 125 kg per hectare. Similarly, they estimated that a reduction of the real interest rate by 10 percent from improvements in the credit system would result in production costs savings equivalent to a yield improvement of 10 kg per hectare. Thus several alternative sources of productivity growth could potentially improve the current and future competitiveness of Kazakhstan's wheat producers.

Rethinking Government's Role in the Wheat System

Kazakhstan's agricultural sector, and its wheat production system in particular, has experienced many difficulties in the last decade. Some level of decline in production and productivity was not unexpected given the fundamental changes occurring in the economy. These have been natural adjustments to the liberalization of the economy from the heavily subsidized production environment created during the Soviet era. Although attitudes and behavior among players in the wheat system have not yet completely shifted to a new paradigm, the realization among many that new directions and priorities need to be incorporated into the research system has begun to take root.

The changes in the Kazakh economy since independence, however, have radically altered the incentive structure within agriculture for farmers. Following the recommendations for the use of machinery and inputs requires both the availability of the inputs and the financial ability to purchase them. By 1993, with the exception of a few selected goods and services (including bread and some agricultural outputs), prices of all retail goods, including the chemical inputs for wheat production, were determined through market mechanisms (Pomfret 1995). The combination of price increases for required agricultural inputs, low output prices for wheat, and high inflation rates during the early years of independence contributed to debt and severe financial constraints on many farms (World Bank 1999). Recent years have witnessed shrinking wheat production resulting from unaffordable and unprofitable technology. Also, unlike the irrigated regions in the south, the northern region has far fewer alternatives to diversification out of wheat. Much remains to be done in reshaping the research, extension, and production system.

The wheat system, however, cannot operate in isolation from the rest of the economy. The research infrastructure and the technology it produces cannot single-handedly be responsible for increasing levels of wheat productivity and quality. It is unrealistic to make changes solely in the wheat system and expect them to bring about the desired improvements in productivity and competitiveness. Much also depends on the creation of production incentives through additional policy measures. The Government of Kazakhstan currently maintains relatively little involvement in influencing wheat production. Investment in research and extension is minimal. Likewise, infrastructural investment—development of roads, terminal facilities, marketing system, and agribusiness promotion—remains very low. The government has divested itself of involvement in the business of input subsidies and wheat procurement with some expectations that private traders and companies will step in to create a private marketing system for grain.

The government's reliance on free market mechanisms and private industry is one thing when markets are fully functioning; the signals transmitted are clear; an effective private industry exists; and overall policy goals in terms of growth, income distribution, and food security are satisfied. However, the economic infrastructure in Kazakhstan has not yet reached the point where competitive market conditions coordinate the smooth operation of the system through signals clearly transmitted by market mechanisms. The Kazakh government would be well served by better defining its role in facilitating improved productivity through some combination of support for research, extension, market and infrastructure development, and development of financial institutions. At the very least, it is in the government's interest to clearly define its objectives for the country's wheat sector and to assess its role in creating the environment necessary to achieve those objectives. Regaining lost levels of productivity can only be accomplished with some level of coordinated policy linking the priorities of the wheat system with those of the agricultural sector and the economy as a whole.

References

Goletti, F., and P. Chabot. 1999. Food policy research for improving the reform of agricultural input and output markets in Central Asia. Paper presented at the Central Asia Workshop on Food, Agriculture, and Natural Resource Policy Research: Setting the Priorities, 19-21 July, Tashkent, Uzbekistan.

Gossen, E. 1998. Agrolandscape, agriculture and forestry management as the basis of sustainable grain production in the steppes of Eurasia. In *Spring wheat in Kazakhstan: Current status and future directions*, eds. Morgounov, A., A. Satybaldin, S. Rajaram, and A. McNab. Mexico City: International Center for Maize and Wheat Improvement (CIMMYT).

Kaskarbayev, Z. A. 1998. Research on spring wheat technology in northern Kazakhstan. In *Spring wheat in Kazakhstan: Current status and future directions*, eds. Morgounov, A., A. Satybaldin, S. Rajaram, and A. McNab. Mexico City: CIMMYT.

Longmire, J. and A. Moldashev. 1999. *Changing competitiveness of the wheat sector of Kazakhstan and sources of future productivity growth*. CIMMYT Economics Paper 99-06. Mexico City: CIMMYT.

Movchan, V. 1998. Spring wheat breeding in Kazakhstan. In *Spring wheat in Kazakhstan: Current status and future directions*, eds. Morgounov, A., A. Satybaldin, S. Rajaram, and A. McNab. Mexico City: CIMMYT.

Pandya-Lorch, Rajul. 1999. Prospects for global food security: A Central Asian context. Paper presented at the Central Asia Workshop on Food, Agriculture, and Natural Resource Policy Research: Setting the Priorities, 19-21 July, Tashkent, Uzbekistan.

Pingali, P. L., and M. W. Rosegrant. 1998. Supplying wheat for Asia's increasingly westernized diets. *American Journal of Agricultural Economics* 80 (5): 954-959.

Pomfret, R. 1995. *The economies of Central Asia*. Princeton: Princeton University Press.

Rosegrant, M. W., M. Agcaoili-Sombilla, and N. D. Perez. 1995. *Global food projections: Implications for investment*. Food, Agriculture, and the Environment Discussion Paper 5. Washington, D.C.: International Food Policy Research Institute.

Sadyk, B., and A. Morgounov. 1999. Ongoing IDF grant activities on improving the wheat sector in Kazakhstan: 1998-99 results. Paper presented at the National Seminar on Agricultural Science and Support Services for Improving the Efficiency of Kazakhstan's Agricultural Sector, November, Astana, Kazakhstan.

Satybaldin, A. 1998. Economic policy for grain production in Kazakhstan. In *Spring wheat in Kazakhstan: Current status and future directions*, eds. Morgounov, A., A. Satybaldin, S. Rajaram, and A. McNab. Mexico City: CIMMYT.

Satybaldin, A., G. A. Kaliev, M. I. Sigarev, and G. A. Nikitina. 1999. National strategy to reform the system of agricultural research, education, and knowledge dissemination. Paper presented at the National Seminar on Agricultural Science and Support Services for Improving the Efficiency of Kazakhstan's Agricultural Sector, November, Astana, Kazakhstan.

Shegebaev, O. S. 1998. Scientific support for spring wheat production in Kazakhstan. In *Spring wheat in Kazakhstan: Current status and future directions*, eds. Morgounov, A., A. Satybaldin, S. Rajaram, and A. McNab. Mexico City: CIMMYT.

Timmer, C. P. 1987. Corn in Indonesia's food policy. In *The corn economy of Indonesia*. Ithaca, N.Y.: Cornell University Press.

World Bank. 1999. *Revival of the seed sector in Kazakhstan*. Washington, D.C.

PART III

Agricultural Policy Reforms—Progress Made and Challenges Ahead

9 Food Policy Reforms for Sustainable Agricultural Development in Uzbekistan

ABDUVOKHID JURAEV, FARRUK AKNAZAROV, ABDULLA FAYZULLAEV, ALISHER TASHMATOV, ROMAN OSPANOV, SHUKHRAT EGAMBERDIEV, HASAN BURIEV, OTABEK SHOMURATOV, SULTONALI MEKHMONOV, DILSHOD HIDIROV, AND RAFSHAN TURGUNBAEV

Given the present conditions of demographic growth and the complicated transition toward a market economy in Uzbekistan, the problems of food security and designing policies to achieve it without compromising economic growth are on the top of the policy agenda. The problem of food security is particularly real and complex in all the countries of the Commonwealth of Independent States (CIS) attempting to liberalize their markets through various reforms.

The food security in any state depends on the fundamental potential of that country's agricultural production; the level of its investment in the agricultural and food sectors, including the agroindustrial system; and the provision of appropriate services to food producers to create the socioeconomic conditions needed for their normal activity. Achieving full harmony between the sustainable use of natural resources and the corresponding food security policy is possible with the sustainable development of agriculture. The rural development triangle comprises three priorities: agricultural development, rational use of the natural resource base, and achievement of food security.

The objective of this chapter is to review food policy and natural resource issues in Uzbekistan in the context of sustainable agricultural development. The chapter is organized into the following sections: trends in Uzbek agriculture during the transition, major food policy reforms, and concluding remarks.

Uzbek Agriculture during Transition

During the transition toward a market economy in the countries of the former Soviet Union (FSU), the government's main role has been to develop a legislative infrastructure as the foundation for private property and the freedom of contractual relations. As a part of this process, the government must modify its actions to complement the changes occurring in the economy. In the course of reforming the financial system of agriculture, outsiders might recommend that the country—which is undergoing a complex transition period—blindly follow the principles of a market economy. A major lesson learned thus far has been that market liberalization and the transformation of people's attitudes toward markets should be developed simultaneously. The role of the government should not be neglected but rather studied and reviewed in a new light.

Since independence, the government has maintained direct control over agriculture. The privatization process has been slow, and farmers continue to rely on governmental control. Nonetheless, the existing mechanisms of governmental control do apply to the development of independent agricultural production. It is necessary to evaluate the impact of privatization on the sustainability and productivity of agricultural production in the short run on transformation of this sector to a market economy in the long run. In practical terms, it would be expedient to diminish the extent of direct governmental control and allow market-regulating mechanisms to determine the supply of and demand for agricultural commodities.

Agriculture occupies an influential position in the economy of the Republic of Uzbekistan. More than 60 percent of the population lives in rural areas, about 40 percent of the adult labor force is engaged in agriculture, and 25 to 30 percent of the gross domestic product (GDP) and 55 percent of foreign currency income flow from agriculture. Every industrial sphere in the republic is related to agriculture. (For basic statistics see Table 9.1.)

This year Uzbekistan is celebrating the eighth anniversary of its independence. Eight years is a rather short time in terms of history, but during this time the state had to reform the entire economy, develop scientifically, and implement the new Uzbek model of reform.

Uzbek Agriculture under the Soviet Union

Under the Soviet regime, Uzbekistan specialized in the production and supply of cotton fiber. The most fertile lands were allocated to cotton. The total cotton-sown area comprised 70 percent of the overall irrigated land area, and in some regions more than 80 percent. As a result, agricultural development was one-sided for many years, and the existing potential to increase food production was used poorly. Sown areas of vegetables, fodder crops, orchards, and grapes constantly decreased.

Basic types of pastry products, potatoes, sugar, livestock, and a number of other foods were imported from other regions of the Soviet Union. The average annual volume of imports included flour and pastry products, 1,700 thousand tons; potatoes, 450 thousand tons; sugar, 300 thousand tons; meat and meat products, 478 thousand tons; and milk and dairy products, 2,929 tons. Uzbekistan even imported certain kinds of fruit.

Naturally, the imported products could not satisfy public demand, in either quality or quantity. In addition, Soviet regulations limited the size of household plots as well as the number of animals rural families could own. These factors complicated an already tense situation in terms of provision of food products to the population.

TABLE 9.1 Basic statistics of Uzbekistan

Independence day	1 September 1991	
Area	447,400 square kilometers	
Administrative divisions	Republic of Karakalpakistan and 12 provinces	
Population	24 million	
Largest cities (population)	Tashkent (2.3 million) Samarkand (385.0 thousand) Namangan (375.1 thousand)	
Ethnic groups (percent)	Uzbek (77) Russian (6) Tadjik (5) Kazak (4) Others (8)	
Basic religion	Islam and Russian Orthodox	
Language	Uzbek (State language)	
Natural resources	natural gas, oil, coal, gold, silver, copper, lead, tungsten, uranium	
GDP structure, 1997 (percent)	Agriculture (31) Industry (27) — Manufacturing (9) Services (42)	
Trade partners (percent)	**Exporting** Russia (15) United Kingdom (10) Switzerland (10) South Korea (7) Belgium (4) Kazakstan (4) Tadjikistan (4)	**Importing** Russia (16) South Korea (11) Germany (8) United States (7) Turkey (6) France (5) Kazakstan (5)
National currency	Som (US$1 = 125 soms)	
Taxes (percent)	Profit tax (35) VAT (20) Income tax (15-45)	

Source: Ministry of Agriculture and Irrigation

Major Food Policy Reforms

After gaining independence, Uzbekistan needed to establish a policy of food provision that took into account the circumstances under the Soviet regime. The first step toward reforming the country's food security policy was enlarging household plots in rural areas and giving plots to those families who had none previously. Starting in 1990, the land holdings of 1.5 million families were enlarged, and more than half a million families received new lands. As a result, these families moved closer to food self-sufficiency and were able to sell some surplus produce in local markets.

The reform of cropping patterns was the second major step toward achieving food security. The Government of Uzbekistan decided to decrease the area sown under cotton while increasing the area allocated to vegetables, fodder crops, orchards, and vineyards. In addition to changing these cropping patterns, the government adopted a special program on grain security which allowed Uzbekistan to stop importing grain and to use locally produced grain.

In the third major policy decision, the government adopted the 1998 Presidential Program for deepening the economic reforms in agriculture, as well as a special regulation for its implementation. These documents built on the legislative objective of deepening the economic reforms by diversifying the type and number of owners and expanding their rights to dispose of and sell their production with no limits. They also envisaged the rational use of land and water resources through the introduction of modern technologies and the expansion of national and foreign investments.

Among the five principles of legal state establishment declared by the President of Uzbekistan, Islam Karimov, the most basic is that the "State is the main reformer and realizer of reforms." Thus far, the legislative base for deepening the agricultural reforms has been established. According to the regulations of the Cabinet of Ministers instituted in March 1998, the program to deepen the reforms in agriculture was adopted to provide sustainable development of the sector. President Karimov noted, "When we evaluate the adoption of this program, we should remember one idea: this program is a guarantee of the sustainable and dynamic implementation of the reforms and the renewal of our society."

Livestock Sector Reforms

During the years of reforms, livestock production has undergone a large shift toward the private sector. About 40 percent of the livestock population was transferred to private ownership, which includes 25 percent of cows and 33 percent of goats. As a result, privately owned livestock increased by 60 percent, including cows by 48 percent, sheep and goats by 48 percent; and the total livestock population—owned by both collective and individual farms—increased by 21 percent.

Reforming State Enterprises

All state enterprises (except those engaged in seed production and breeding activities) are now privatized. Today, 92 percent of all agricultural commodities are produced in the nongovernmental sector. The introduction of free market prices and the stimulation of the production process are proceeding smoothly. Overall, the reform process encompasses liberalizing and decentralizing production structures, establishing private *dekhkan* enterprises (regulated small scale farms), privatizing livestock farms, and creating a stratum of owners in the rural areas who have long-term tenure of fixed-land plots and can sell their produce.

Production of Agricultural Commodities

After independence, the supply deficit of essentials like meat, milk, fruit, and vegetables was cut significantly by establishing smallholdings and providing favorable conditions for farmers. Surpluses are now sold in the city and *dekhkan* markets. The increases in agricultural production resulting from these liberalization policies are shown in Table 9.2 and Figure 9.1. Some growth also occurred

TABLE 9.2 Production of agricultural commodities in Uzbekistan, 1994-98

Commodity	1994	1995	1996	1997	1998
	(thousands of metric tons)				
Raw cotton	3,938	3,934	3,350	3,600	3,206
Wheat	1,363	2,347	2,742	3,073	3,555
Rice	499	328	450	386	346
Potato	567	440	514	692	691
Tomato	488	427	269	250	320
Cucumbers	8	8	6	6	8
Other vegetables	603	548	494	456	399
Melons	578	472	470	376	470
Fruit	555	602	605	547	544
Beef	390	391	362	386	393
Lamb meat	73	83	75	63	63
Poultry meat	24	16	12	10	11
Milk	3,731	3,665	3,403	3,406	3,495
Eggs (millions)	1,574	1,232	1,057	1,075	1,165
Dry fruit	16	13	11	8	10
Vegetable oil	361	340	272	276	279
Sugar	0	0	0	0	0

SOURCE: Ministry of Agriculture and Irrigation.

FIGURE 9.1 Yield of agricultural crops in Uzbekistan, 1994–98

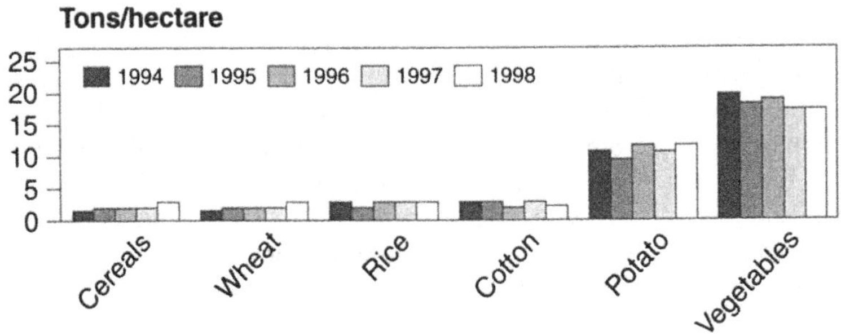

Source: Author's calculations.

in the production of food products not regulated by the government, like potatoes, tomatoes, vegetables, beef, milk, and eggs.

Uzbekistan has also paid significant attention to grain independence and to the supply of essential food products and raw materials to the food industry. As a result, since 1996, the volume of grain imports has dropped significantly.

Policies for achieving self-sufficiency in strategic products are of great importance, but at the same time, they can undermine overall food security. The production of strategic commodities (which are economically unprofitable) requires large investments and influences both export incomes and the import of other food stuffs, inputs, spare parts, and new technologies. The solution to this dilemma largely depends on the priorities set by the country.

Consumption of Food Products

Achieving food self-sufficiency to meet the recommended daily allowances of nutrients for all the population has been one of the major objectives of the agricultural sector reforms. Since 1990, the consumption of food products per capita has decreased in Uzbekistan, largely because of the problems of the transition period and the breaking of economic and trade relations with neighboring countries in Central Asia. Some growth in consumption has appeared since 1996. However, the consumption of some food products still falls below nutritional norms. In some cases, shortfalls are made up by increased consumption of bread, flour, and pastry products, which are higher in calories (Table 9.3, Figure 9.2).

One more indicator of the food security problem is the share of expenditures for food products out of total income. This indicator has decreased since 1997, indicating a positive movement toward solving this problem.

Reforms in the Agricultural Processing Sector

Issues of agricultural food product processing and preservation occupy an important place in the achievement of food security. Building a network of processing

TABLE 9.3 Per capita consumption of food products in Uzbekistan, 1994-98

Food product	Annual consumption norm	1994	1995	1996	1997	1998
			(kilograms)			
Meat and meat products	75.6	33.0	33.0	30.0	31.0	31.0
Milk and diary products	196.2	173.0	160.0	149.0	149.0	153.0
Eggs (pieces)	180.0	63.0	53.0	45.0	46.0	50.0
Fish and fish products	1.6	1.4	1.4	1.2	1.5	1.8
Bread	186.6	152.0	157.0	183.0	201.0	198.0
Potatoes	150.0	25.0	23.0	27.0	36.0	36.0
Vegetables and melons	84.0	130.0	120.0	102.0	90.0	99.0
Fruit, berries, and grapes	24.0	24.0	29.0	29.0	26.0	26.0
Sugar	25.2	9.0	9.9	14.0	13.2	18.1
Vegetable oil	13.8	13.5	13.9	13.3	13.5	13.6
Share of food product costs in total income (percent)		68.9	69.4	68.6	67.1	64.5

SOURCE: Ministry of Agriculture and Irrigation.

enterprises intended for agricultural production and preservation will increase the capacity for processing and preservation and will reduce the losses from pests and disease and through harvest and transport.

Unfortunately, the development of this sector has far to go. The costs of marketing, processing, and storage are in some cases several times higher than the costs of production, leading to an incorrect distribution of incomes and negatively influencing the investment flow structure. Consumption from own production prevails, as state marketing channels are not commercially sufficient and bazaars (local markets) are the only channels available to small farmers. Further development of the processing sector is one possible solution to this problem, which could be achieved through the purchase of mini-plants. But for people living in rural areas, their lack of financial resources and shortage of experience and information, coupled with high interest rates at commercial banks, offers little hope in this direction.

Imports and Exports under the Reforms

In the structure of exports, the largest share is occupied by raw material production, in particular, cotton fiber (Figure 9.3). The same can be seen in the structure of investment. More attention must be paid to developing the processing industry as the prices of raw materials continue to decrease.

FIGURE 9.2 Consumption of food products in Uzbekistan, 1994–98

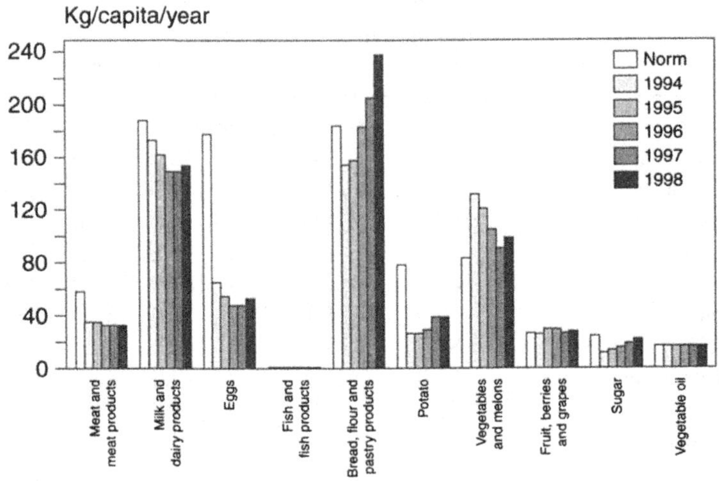

Source: Author's calculations

FIGURE 9.3 Commodities exported from Uzbekistan, 1998

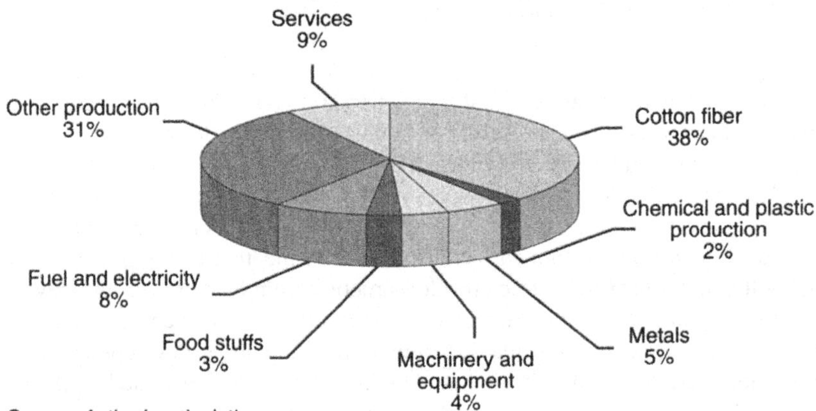

Source: Author's calculations

With regard to imports, the status of certain goods in consumer markets is a subject of concern. Because of increased trade restrictions, the trend has been to rely mostly on the local production of food. However, decreases in the import of some goods reflect a drop in their supply because domestic production has not been able to make up the difference (Table 9.4). For example, the supply of flour, vegetable oil, and cheese has declined as their imports have declined. These declines are further complicated by the increasing demand for such products from the growing population. For some products, such as meat, potatoes, and tobacco,

TABLE 9.4 Fluctuation of the import share of the consumer market, 1997 and 1998

Commodity	1997			1998		
	Supply	Including import	Import share (%)	Supply	Including import	Import share (%)
Meat (thousands of tons)	189.7	36.2	19.1	228.0	6.7	2.9
Flour (thousands of tons)	1237.4	103.7	8.4	1,045.9	71.7	6.9
Vegetable oil (thousands of tons)	118.0	46.9	39.7	112.2	27.4	24.4
Fat (thousands of tons)	14.7	8.8	59.9	11.9	10.3	86.6
Cheese (thousands of tons)	2.7	1.1	40.7	2.4	0.6	25.0
Sugar (thousands of tons)	172.8	172.8	100.0	170.0	170.0	100.0
Potatoes (thousands of tons)	476.3	21.6	4.5	542.5	8.3	1.5
Textiles (billions of soms)	14.3	0.6	4.2	21.3	0.5	2.3
Shoes (millions of pairs)	18.5	6.1	33.0	14.7	12.0	81.6
Tobacco (billions of soms)	8.9	0.1	1.1	17.4	0.08	0.5
TV sets (thousands of pieces)	123.3	21.8	17.7	55.9	5.9	10.6

SOURCE: Ministry of Macroeconomics and Statistics.

the supply has increased while their imports have dropped. Uzbekistan continues to import all of its sugar supply.

Rural Unemployment under the Reforms

As shown in Table 9.5, the growth of unemployment currently outstrips the growth of the population available for work. Unless the rural population displaced from agriculture is able to find employment, such a trend will continue at an even higher rate.

The Program of Farm Restructuring stated that, by the year 2000, state farms would be restructured into agricultural cooperatives on the basis of shares. Currently, 65 cooperatives have been established out of the 1,802 state farms, resulting in the layoff of 28 percent of the state farm employees. If that proportion remains constant as the remaining farms are restructured, about 500,000 rural workers will join the ranks of the unemployed in the next two to three years. This would represent about 10 percent of the total number employed in agriculture.

Rural Financial Reforms

One of the major preconditions for economic growth is an effectively functioning finance system. Several attempts have been made to reform the existing agricultural finance system. A new finance-credit enterprise, Tadbirkorbank, was established to provide financing to emerging dekhkan and small farms.

TABLE 9.5 Population and employment, 1997 and 1998

	1997	1998	Growth
	(thousands)		(percent)
Population	23,560.0	23,966.5	1.73
Labor force	11,805.4	11,998.9	1.64
Economically active population	8,645.2	8,840.1	2.25
Employed	8,694.8	8,800.0	1.21
Unemployed	35.4	40.1	13.28

SOURCE: Ministry of Macroeconomic and Statistics.

Tadbirkorbank is also in charge of granting credits from the Small and Medium Businesses Development Promotion Fund. The Agroprombank was privatized and divided into three separate banks specializing in different crops. However, these measures have not achieved their objectives, and the existing situation still depends on a system of centralized financing. Banks have not yet begun to engage in internal banking reform.

One of the major problems in agricultural finance is an emerging disproportion caused by the different rates of liberalization in some sectors of activities. For instance, commercial banks can easily make transactions in the equity market, whereas an enterprise can open only one bank account (this is done for the sake of controlling tax payments). Outdated instructions and regulations are applied to new conditions, resulting in improper use of financial resources and investment in nonproductive activities rather than effective use of credit. These practices leave banks little incentive to attract deposits and enlarge their volumes. The current practices absorb the existing resources and negatively influence the banks' potential as financial mediators and providers for further development of the sector.

For the agricultural finance system to function properly, three goals must be achieved: make commercial crediting independent, improve the attractiveness of deposits, and diversify financial resources. As a first step, financial development agencies must get financing from both the state and international sources for the development of private financial activity.

Commercial banks must resume their main function as financial mediators. Banking processes should be studied; and banking personnel should be trained to strengthen banking potential, analyze credit ability, better monitor issued credits, and increase the number of banking services. For example, to strengthen bank potential, some negative practices must be abolished, including certain kinds of controls that are limiting the liquidity of deposits and promoting unfair competition. Having improved capitalization, banks must end their dependence on the system of centralized crediting. Furthermore, banks must develop the potential to issue medium-term and long-term credit.

The farm restructuring process should be accelerated to achieve the agricultural policy goals. Farms should be released from constraints on their production, and incentives for improving production should be strengthened. In addition, outdated equipment and techniques should be replaced, and a service system of input resource supply should be established.

Reforming the agricultural finance system is the main component of transforming agriculture into a sector of the market economy. Reform means more than simply providing the sector with the necessary resources. Reform means giving farm leaders access to credit on the basis of business acumen and responsibility, and allowing them to purchase inputs in a free market and to achieve a profitability level sufficient to repay borrowed credit and to accumulate capital. Such transformations will make possible a significant increase in agricultural productivity.

Reform of the banking sector is an important part of the general reform. Will the activity of specialized banks affect the process of agricultural reform implementation? How can accessibility to bank credits be improved without risking a loan default crisis? Answers to these and many other questions are still needed.

Sustainable Natural Resource Use under the Reforms

Environmental issues are also of major importance for Uzbekistan. Their solution will determine the living conditions for present and future generations and will affect the development of all sectors of the economy.

The Republic of Uzbekistan spans 45 million hectares. Agricultural production occurs on 33.2 million hectares, of which 4.2 million hectares are irrigated. Land degradation due to soil salinity in irrigated areas continues to be a major environmental problem in Uzbekistan. During 1975-85, reclamation of saline lands was especially intensive. About 1 million hectares were reclaimed, and the irrigated area was extended by 1.4 times. However, the intensive reclamation of rather saline and poorly irrigated lands increased saline areas by 0.8 times over the last 15-20 years. At present, salinity affects 2 million hectares, out of which 850,000 hectares are medium or heavily salted. Almost all of the territory of main Karakalpakistan (90-95 percent) is totally salted; and in Bukhara, Syrdaria, Kashkadaria, and Khorezm provinces 60-70 percent of the land is salted. In 1991 intensive land reclamation was terminated in the republic because of the extreme amount of irrigation required and because of the problems facing the Aral Sea.

The environmental problems in irrigated areas are escalating because of a lack of financial resources. According to a World Bank assessment, to restore and maintain the irrigation balance at a satisfactory level, it will be necessary to invest US$350 per hectare of irrigated land, a level 20 times greater than existing financing. The main consequences of the low financing of irrigation activities are as follows:

- Silting of channels and reducing their outlet ability
- Increasing water losses from drainage and filtration channels and decreasing irrigation network efficiency
- Decreasing efficiency of existing drainage systems
- Decreasing water supply for irrigation
- Accelerated salinization of arable lands

The failure to address these problems will contribute to the worsening of the condition of irrigation and drainage systems, and will eventually decrease crop yields. For instance, if the salinity of 1 percent of the land increases from low to medium, the yield losses will be about US$12.5 million per year. The humus content of soils—the main indicator of soil fertility—has decreased by 30-50 percent. Forty percent of lands in irrigated areas are characterized by low and very low humus content (0.4-1.0 percent). The total area of low-productive arable lands (including heavily salted, gypsous, eroded, and stony soils) is 500,000 hectares. Low-productive arable land is present in almost all provinces, but the largest portions are concentrated in Kashkadaria province (134,000 hectares), Karakalpakistan province (70,000 hectares), and Samarkand province (60,000 hectares). Water and wind erosion and range degradation are also important issues for agriculture.

Concluding Remarks

In conclusion, because of the economic, geographic, demographic, cultural, and other differences among the Central Asian republics, there is no unified formula for the achievement of sustainable agricultural development and the resolution of their food security problems. In Uzbekistan these issues require in-depth research and analysis. We appreciate the efforts of the International Food Policy Research Institute (IFPRI) in organizing the present workshop, which is extremely valuable at the current stage of our country's development. We look forward to future collaboration with IFPRI and other international organizations to improve our knowledge of agriculture, food security, and sustainable environmental development, and to improve the prosperity of the populations in our republic and in the other Central Asian countries.

10 Policy Reforms and Food Security in Uzbekistan

RASULMAT KHUSANOV

One of the important issues addressed by this workshop is the problem of how a country provides its people with a sufficient food supply. Achieving food security is an important condition for establishing a stable and safe atmosphere in any country and guaranteeing an effective economy. This problem is an integral part of a country's overall socioeconomic status and is closely tied to demographic growth and improving ecological processes.

In this modern era, as the countries of the Commonwealth of Independent States (CIS) enter into market-driven economics, the problem of food security is especially crucial. The provision of food security in any state depends on

- the basic potential for agricultural production;
- investment in the food production sectors, including the agroindustrial system;
- the provision of appropriate services to food producers; and
- the establishment of essential socioeconomic conditions for normal activity.

Taking these factors into account and before talking about policy aspects of food security, I would like to discuss the basic potential for agricultural production in Uzbekistan. Uzbekistan is celebrating the eighth anniversary of its independence this year. This is a rather short time in terms of history, but in this period, the State has had to reform the entire economy, to advance scientifically, and to develop a practically new Uzbek model of reforms. Uzbekistan has had to establish its food production system anew because, within the former Soviet Union's specialized and concentrated production process, Uzbekistan had focused primarily on the agricultural production and supply of cotton.

The most fertile lands were allocated to cotton. Cotton production covered 70 percent of all irrigated lands, and in some regions more than 80 percent. As a result, agriculture developed one-sidedly for many years, and the potential to increase food production was used very poorly. Sown areas for vegetables, fodder crops, orchards, and grapes fell constantly. Basic types of pastry products, potatoes, sugar, livestock, and a number of other foods were imported from other Soviet regions. Norms regarding diversity of diets were not followed, and pastry products (such as wheat bread) provided the greatest caloric value of the food diet. In the former Soviet Union, the largest volume of pastry product consumption per capita was in the Central Asian republics.

The annual average volume of imports was as follows: flour and pastry products, 1.7 million tons; potatoes, 450,000 tons; sugar, 300,000 tons; meat and meat products, 477,800 tons; milk and diary products, 3,000 tons. The irony of the situation became clear when certain species of fruit had to be imported to Uzbekistan.

Naturally, the imported products satisfied public demand in neither quality nor quantity. Besides this, Soviet regulations limited the size of household plots and the number of animals owned by rural families. This complicated the already difficult situation in terms of the provision of food products to the population.

After independence, it was necessary to reform the overall policy of food security for the country. The first step in this direction was to enlarge household plot size in rural areas and to give plots to those families who had none previously. By presidential decree, from 1990 to 1991, the plots of 1.5 million families were enlarged from 0.18 hectare to 0.25 hectare, and 580,000 families were provided with land. All in all, during this short period, the total area allocated for households was 291,500 hectares of irrigated lands. Thus, the total area under households was doubled to 516,000 hectares.

These increases resulted in some improvement in food provision to the population. Many rural families began to sell some surplus in the market, meaning that they turned from consumers into net producers. One should note that the allocation of land to rural families was occuring in the context of decreasing irrigated area per capita. That is, in 1960, there was 0.25 hectare of arable irrigated land per citizen in Uzbekistan; in 1970 that dropped to 0.20 hectare; in 1980, to 0.19 hectare; and in 1990, to 0.16 hectare. By the year 2000, the availability will be no more than 0.10 hectare per capita.

The decision to revise cropping patterns was an important step toward food security. Cotton-sown areas decreased by 1.5 million hectares while the volume of production rose through increases in yield. The released area was allocated to vegetables, fodder crops, orchards, and grapes. Specific attention was given to increasing the areas sown under grain crops. Irrigated wheat areas rose to 1.3 million hectares. A special program on grain security was also adopted.

The decision to increase grain production allowed Uzbekistan to stop importing grain and to provide itself with locally produced grain. It was a timely and bold political and economic decision, which continues to prove its worth. Uzbekistan is a self-sufficient grain producer, and the implementation of the Program on Grain Independence has allowed Uzbekistan to turn in a short period of time from a grain importer into a net grain exporter.

The food policy of Uzbekistan also envisaged self-provision of such food products as potatoes, rice, fruits, and vegetables, and envisaged increasing production of sugar and livestock. In large part, those tasks are being resolved successfully. Thus, Uzbekistan provides itself with potatoes, rice, basic fruits, grapes, and vegetables through its own production. A new sector is being established to produce sugar. Sugar beet planting areas have been enlarged, and several sugar producing plants have been established as well. Hence, the use of existing potential has allowed Uzbekistan to strengthen its food security and minimize its dependence on other countries.

Another important aspect of food security is the provision of livestock and cooking oil. During the Soviet regime, these products were imported from other regions, and local production was at a low level. Since independence, a reform of the livestock sector was carried out. Some livestock farms were privatized and others were sold to private owners through auctions. Today, every rural family has its own cattle, which, along with providing products for home use, can supply the market with these products as well.

During privatization, some orchards and grape farms were also transferred to private ownership. Now, fruit and vegetable production is entirely assigned to the private sector. State orders for agricultural production were abolished for all types of produce except cotton and grain, enabling food producers to choose which products to produce, in which markets to sell, and at what prices.

Agriculture provides an example of the reforms ongoing in all spheres of the economy. During the reform process, an appropriate legal basis was developed to accommodate the different tenure forms. All of these forms are based on goods and food production. A mechanism of balanced agricultural development is being created to provide not only social stability but also economic independence for the country. These are the basic components of Uzbekistan's food security policy.

The most prominent example of small-scale farming is *dekhkan* farms. These were established to take the place of the former household farms but are quite different from household farms. A special law on *dekhkan* farms was adopted wherein all rights and responsibilities are stipulated by appropriate regulations. In addition to growing produce for their own needs, *dekhkan* farms can obtain extra land plots to produce food for sale. They have a right to function as both physical and legal entities, and banks can provide them with credit. At the moment, there are more than 3 million *dekhkan* farms in the republic. These farms produce 53 percent of the vegetables, 68 percent of potatoes, 59 percent of fruits and vegetables, 37 percent of grapes, 89 percent of meat, and 92 percent of milk in the country. As shown by these figures, *dekhkan* farms produce the main share of agricultural production in Uzbekistan. For this reason, the government is dedicated to establishing the conditions necessary to support these farms' effective operation. At the same time, this form of farming should not be considered representative. *Dekhkan* farms have been the main producers during the transition period, but in the future other forms of farming are to be established for the following reasons:

- First, small land plots do not allow the application of high productivity modern technology and techniques, therefore leading to high costs;
- Second, poor breeding activity and seed production and the use of technologies mainly oriented to manual labor do not permit increases in production quality. Therefore, *dekhkan* farms will not be able to contend with high market competition; and
- Third, selling most production to local consumers excludes the possibility of export development.

Nevertheless, the *dekhkan* farms will remain the basic food suppliers during the transition period.

Individual farms are the medium-scale level of farming. The law "On farm enterprise" states the rights and responsibilities of these farms. Individual farms are bigger in terms of production volume compared with *dekhkan* farms. At the moment, there are more than 20,000 operating individual farms. Most of them specialize in fruit and vegetables and livestock production. The number of farmers engaged in crop production, including grain and cotton production, is increasing. Unlike *dekhkan* farms, individual farms possess larger land plots and better means of production. They can practice more intensive farming with lower costs through the use of mechanization and fertilizers. Having their own transport, these farmers can choose an appropriate market and export a part of their production outside their local area. Although the share of individual farms does not exceed 10 percent of the total volume of production, I believe they are a more ideal form of farming than the *dekhkan* farms and large agricultural cooperatives (*shirkats*).

In particular, the advantages of the individual farms include the following:

- The problem of ownership is resolved. When a farmer gets land with a long-term tenure, he is the master of production and can dispose of it as he wishes.
- The larger scope of production allows for increasing capital efficiency and lower costs.
- A farmer can specialize in the production of a single type of product, which allows him to improve its quality and to be market competitive.
- And a farmer can establish processing enterprises and reap the benefits of vertical integration.

For these reasons, the share of individual farms in food production will continually increase.

At present, as part of the agricultural policy aimed at reforming the agricultural sector, unprofitable cooperative farms are being transformed into individual farms. A large-scale form of farming is the agricultural cooperative or *shirkat*. All legal, economic, financial, and organizational aspects of the *shirkat's* function are stipulated in the law "On agricultural cooperatives (*shirkats*)." These cooperatives focus on the production of cotton, rice, and cereals. More than 80 percent of grain and 90 percent of rice in Uzbekistan are produced by agricultural cooperatives. The large scale of agricultural cooperatives allows them to use production techniques more effectively, to establish seed production and breeding activities, and to use land and labor resources more efficiently. Their size provides them with more opportunities to produce with lower costs and also to establish processing and conservation activities. However, the problems of property rights and ownership are not yet fully resolved. Some attempts are being made to resolve these issues by establishing family subcontract work with distribution of land ownership on a shared basis. However, some problems relating to ownership remain. The organizational structure of agricultural cooperatives still needs improvement. Nevertheless, agricultural cooperatives will remain the main producers of basic food crops like wheat, rice, and livestock in the near future.

The realization of these objectives will require the integration of several measures. Particular attention should be paid to the following: external regulation; arrangement of trade and investment contracts; and participation in interstate and international organizations. In addition, Uzbekistan, with its large population should be at the forefront of the Central Asian republics in dealing with issues of food security.

Forming a network of processing enterprises intended for agricultural production and providing for its long-term maintenance will play an important role in achieving food security. For example, many kinds of produce are perishable and difficult to transport. Research has shown that overall losses of fruits and vegetables caused by pests and diseases and during transport (as well as in other stages of production) represent 35 to 40 percent of the total volume produced. Therefore, the government is implementing the necessary policy measures to promote appropriate processing and conservation practices.

Since independence, the capacity of processing enterprises has increased 1.4-1.6 times. Most of these enterprises are privatized; and farmers and agricultural cooperatives have established many small processing enterprises. These enterprises will decrease losses and improve the supply of agricultural produce.

These approaches to reform form the basis of Uzbekistan's economic and social orientation, benefiting from mechanisms of the free market. In this way, the state policy aims to achieve its priorities of providing the population with food products, eliminating malnutrition, and establishing the conditions necessary for normal and long life. To achieve these goals, the President has issued a decree ensuring the availability of nine essential food products at all retail outlets, with price monitoring to avoid price increases.

These are the basic political aspects of food security in Uzbekistan. I would like to draw attention to one more problem. Uzbekistan possesses soil-climatic conditions favorable for growing many kinds of agricultural products on a year-round basis. The potential for production significantly exceeds the republic's needs. Thus, Uzbekistan can export excess production in both fresh and processed forms. This was done in the past. After the Soviet Union's collapse, export levels decreased because of the complexity of customs procedures and transportation, causing significant damage to Uzbekistan's economy and hampering the food security of its neighbors.

I believe it would be appropriate, within the framework of this workshop, to consider the possibility of conducting research regarding the exchange of food products among the Central Asian countries and Russia. We should consider export and import food production balances and make suggestions to the governments of these countries. In terms of scientific potential, the Institute of Agricultural Policy Reforms is prepared to perform such activities in collaboration with other scientists of Central Asia. The export of food products is to become one of the sources of hard currency income to all participating countries and will simultaneously increase the livelihood of farms.

In summary, since independence, the priority of agrarian reform in Uzbekistan was first to diversify agriculture, then to liquidate cotton monocrop production, to improve self-sufficiency of food production, and to develop the processing branch of the agricultural sector. These are the basic aspects of the food security achievements of Uzbekistan.

11 Food Security and Policy Priority Tasks in Uzbekistan

ELEONORA GAZIYANTS

For the past decade, the government of the Republic of Uzbekistan has chosen a course of in-depth, socially oriented economic reform, with the first priority given to the diversification of agriculture. This decision was made because agriculture accounts for 70 percent of internal trade, generates 55 percent of hard currency income, and produces 85 percent of the food products needed to satisfy public demand.

The provision of food security through local means is a factor determining social progress and development in any state. During the Soviet period, food security issues were resolved according to the all-Union system of labor division and specialization. Uzbekistan's role was as the main supplier of cotton, producing 64 percent of the all-Union total. Thus, the majority of livestock and nontraditional products for the region had to be imported.

According to statistical data from the Ministry of Agriculture and Irrigation, at the end of the 1980s, the Republic of Uzbekistan was importing 149,000 tons of meat and meat products; 718,000 tons of milk and dairy products; 160 million eggs; 4.1 million tons of grain; 355,000 tons of potatoes; and 405,000 tons of sugar. Thus, a state of overall food self-sufficiency did not exist in Uzbekistan. At the end of the 1980s, grain production per capita was 83 kilograms (kg) per year; potato production, 13.4 kg, meat products, 21.4 kg; diary products, 135.5 kg; and eggs, 108 units. Expressed as a share of total food demand, Uzbekistan produced 52 percent of its grain needs and 28 percent of potato needs.

Following independence, however, Uzbekistan had to establish a new system to achieve food self-sufficiency. In this regard, the first tasks of agrarian reform were to eliminate the monocrop production of cotton, to achieve grain independence, to develop the processing sector, and to strengthen the country's ability to provide food products to the population.

As a result of policy changes, the importation of grain during 1990-98 decreased to an annual average of 1.6 million tons, while per capita grain and potato production increased 1.9 and 2.2 times, respectively. Production of other products is approaching the self-sufficiency level. However, given the conditions in Uzbekistan—high population growth and limited land and water resources—the consumption of meat, dairy products, eggs, and sugar continues to be less than the recommended levels.

One policy question is urgent and persistent, namely, "Why should we produce grain if we can import it?" Experience has shown that following the elimination of cotton as a monocrop, some changes in cropping patterns began to appear. These changes included increased grain, potato, and sugar beet production; increased soil fertility in irrigated areas; and the achievement of food security through effective agricultural production.

According to estimates of specialists, the potential profit from one hectare of grain (assuming an average yield of 2.1 tons per hectare) is more than US$100. On farms where the yield increases to 5-7 tons per hectare, the profit increases proportionally. Therefore, escaping grain dependence can increase economic growth and improve the welfare of farmers.

These statistics explain why extensive grain production should be replaced by intensive production, including the concentration and specialization of production. With this shift about 500,000 hectares of irrigated lands producing less than 4.5 tons per hectare would be released from grain production and could be used to grow fodder crops for livestock, thus increasing meat and dairy security for the country. The shift is also justified by the fact that the import of grain from other countries is economically unprofitable because of high customs fees and the disruption of transport systems between former Soviet republics.

Some income groups in the population are experiencing difficulties because of the current mixed economic system. Still they clearly realize that their well-being depends on their participation in the new economic system. Such difficulties should not cause too much concern. They are a regular part of the process of political and economic liberalization in countries in transition. Indeed, the country has taken some steps toward building a social safety net while enhancing the process of economic reform.

In addition, citizens themselves will be able to put to use their experience, their individual capabilities, and their knowledge, and choose what income-generating activities to pursue. They might choose either individual activities or entrepreneurship and business, or a combination of public activity and household production, or training and human capacity building. As experience with developed countries shows, many families combine farm activities with other sources of income. In the United States, agriculture is the sole source of income for 45 percent of tenants and 32 percent of hired farm labor. In all other countries, two-thirds derive their sole income from agriculture. Dividing the employed population of Uzbekistan by the number of income sources shows that 32 percent have two income sources. Thus, according to the statistics, each working person is supporting at least two or three family members, usually children or retired parents.

The rather narrow sphere of activities in rural areas, with a limited range of professions and specialties, means that some social strata of the population are not able to meet their income needs. Furthermore, the poor development of rural infrastructure and processing industries leads to unemployment. Finally, the process of privatization has transformed rural enterprises and resulted in the loss of jobs. All these factors create an imbalance between employment opportunities and labor resources.

Thus, to further address the problem of food security and achieve rational and efficient employment, integrated activities must be developed. Both a program for increasing agricultural production and a program expanding the spheres of material and nonmaterial production are needed.

In this regard, scientists and economists face the following important tasks:

- Developing an integrated program to benefit local agricultural producers under conditions of multiple crop production systems with different farm sizes
- Improving the social and production infrastructure in rural areas on the basis of systemized territorial and state regulation, taking into account the specific circumstances of the various regions and the possibilities for integrating the regions into a unified economic area
- Attracting foreign investment to provide jobs, thereby balancing labor resources through increased employment and improved buying capacity of the population
- Improving activities of the regional labor employment services to reduce unemployment among the able-bodied population, particularly among youth

Ultimately, a favorable economic policy environment will be established that will increase the supply of food to the population on the basis of effective employment and growth in real incomes.

12 Attaining Food Security in the Kyrgyz Republic through the Rational Use of Natural Resources

NAZARBEK IBRAGIMOV AND A. J. ASANALIYEV

Because of its geographic location, the Kyrgyz Republic has various and contrasting climates and natural resources. Even though the republic's territory is large relative to its population, the share of land suitable for agricultural use is not significant. Out of a total territory of 19.8 million hectares, around 10 million hectares are arable and only 1.3 million hectares are used for agriculture. Thus, per capita land availability is 0.30 hectare of arable land, of which 0.22 hectare is irrigated.

The Kyrgyz Republic agricultural lands are located in valleys, intermountain cavities, and high mountains, covering elevations between 500 and 3,200 meters above sea level. The temperature varies between 22 and 49 degrees Celsius, and the average annual precipitation ranges between 80 and 700 millimeters per year, depending on the agroclimatic zone (Table 12.1). In this regard, the vegetation diversity of the country is rich. The agricultural zones of the Kyrgyz Republic are so diverse that they represent most of the agroclimatic zones and soil diversities of the world. The republic is also rich in various types of cultivated crops and domestic animals.

Agricultural Production during the Soviet Era

Agricultural production during the years of the Soviet Union followed three categories of property ownership: private property (personal plots of land, the orchard-growing communities), collective-cooperative property (cooperatives and *kolhozes*), and state property (*sovhozes*, pedigreed stock farms, procurement and processing enterprises, and the trade system). However, the collective and state farms and enterprises played the largest role in the production of the main agricultural products.

During the Soviet period, the field and farm workers of the Kyrgyz Republic achieved significantly high yields from their agricultural crops and high productivity from their livestock by applying scientific agricultural practices and livestock breeding. The average yield of grain crops in the republic during the 1980s reached 34.5 tons per hectare; raw cotton reached 28.5 tons per hectare; aromatic tobacco leaves, 2.8 tons per hectare; and sweet beets, 40.0 tons per hectare. The average yield of vegetables was 21.2 tons per hectare, and the average yield of

TABLE 12.1 Agroclimatic characteristics of Kyrgyz Republic agricultural zones

Agricultural zone (area)	Altitude above sea level	Annual precipitation	Vegetation period
	(meters)	(millimeters)	(days)
Irrigated valley agricultural zones: Aravan, Nooken, Bazar-Korgon, Kara-Suu, Suzak, Manas, Panfilov, Jayil, Moscva, Alamedin, Sokuluk, Kant, Isik-Ata, Chui, Kadamjay, Leilek, Batken *rayons*	500-800	300-500	197-212
Mountainous irrigated and pasture land zones: Aksy, Ala-Buka, Kadamjay, Toktogul, Toguz-Toro, Suzak, Kara-Suu, Ezgen, Talas, Bakay-Ata, Kara-Bursa, Isik-Ata, Alamedin, Sokuluk, Moscva, Jayil, Panfilov, Kemin, Leilek, Batken *rayons*	800-1,600	200-600	114-200
Mountainous, intermountainous irrigated, and pasture land zones: Toktogul, Toguz-Toro, Kara-Bura, Chatkal, Suzak, Aksi, Ala-Buka, Kara-Suu, Isik-Ata, Ak-Suu, Tok, Tup, Jeti-Oguz, Kochkos, Tyan-Shyan, Jumgal, At-Bashi, Kemin and Issik-Kul *rayons*	1,600-2,200	200-400	110-170
High-altitude irrigated and pasture land zones: Alay, Kara-Shoro, Suusamir, Alaykul, At-Bashinksky *rayons*	2,200 and more	300-700	60-120

SOURCE: National Statistics Committee.

NOTE: Individual *rayons* have multiple agricultural zones.

fodder crops was as high as 7.0 tons per hectare. More advanced farms managed to achieve 5.0 tons of grain crops per hectare, 15.0 tons of hay per hectare, and 35.0 tons of potatoes and vegetables per hectare.

These achievements in crop production were accompanied by similar developments in the livestock sector. The stock of sheep and goats reached 10 million by 1990, cattle stock reached 1.2 million, the number of horses was 3.1 million, poultry stock reached 13.9 million, and the number of pigs reached 393,000. During the winter, agricultural animals were kept in farm structures. On average, the fertility of sheep and goats was 98-100 lambs per 100 female parents; 100 female horses gave birth to 61 foals; and 85 calves were born to 100 cows. The highest production parameters were those for pig breeding and poultry breeding. These two sectors were fully transformed to run as an industry. During this period, the Kyrgyz Republic held the record among the Soviet republics in the growth of sheep and production of wool clippings.

During the high-yielding years in the Kyrgyz Republic, an average yield of more than 3,300 liters of milk per cow was obtained. Similarly, an average of 3.2 kilograms of wool was clipped from each sheep. Cow milking and wool clipping were fully automated. In this way, the annual increase in the gross agricultural output during 1980-90 was 3.5 to 5.0 percent. The procurement and processing sectors of agriculture also grew at a similar rate. The gross output of the agroindustrial complex was more than 60 percent of the gross national product of the republic. Measures aimed at improving the infrastructure of the agroindustrial complex were undertaken year after year. Although the Kyrgyz Republic was once an economically backward country, it was able to develop an agroindustrial complex during Soviet times, and ranked highly among the15 countries of the former Soviet Union. In many ways, the Soviet Union was essential to the cultural and economic regeneration of the Kyrgyz Republic. Each *kolhoz* and *sovhoz* had its schools, cultural, and community facilities centers. Each farm, on average, had more than 40 specialists engaged in agricultural training.

However, during the last years of the Soviet regime (1980-90), each incremental increase in production came at a high cost. Farms became unprofitable. The decisions made by the Soviet government to improve the situation in agriculture brought few results.

The collapse of the Soviet Union and the formation of independent governments gave the people of these countries political freedom and feelings of self-respect. At the same time it cut all the threads to the single government. This happened without any political or economic relations between the former republics and the other countries of the world. Prices changed, unusual customs houses emerged, and the newly sovereign countries became hostile toward one another. Furthermore, after the collapse of the Soviet Union, the republics that were endowed with energy resources and had accommodated the priority sectors of the national economy remained in an advantageous position, while other republics' sectors did not fare as well.

Policy Reforms and Their Negative Consequences

The transition to a market-based system most seriously damaged the agroindustrial complex in the Kyrgyz Republic. In the agroindustrial complex, decreases in output from the crop and livestock sectors were accompanied by the breakup of the relations between these sectors and the processing enterprises and trade networks. These facts beget the question: "Did the politicians and economists of the Central Asian countries act correctly in their actions following independence?" In our opinion, the politicians and the economists of the Russian Soviet Federative Socialistic Republic should have considered the destiny of the new countries that emerged as a result of the collapse of the Soviet Union and given each the opportunity to create its own infrastructure and to improve on the former economic system. Instead, having fallen victim to great-power chauvinism, they cynically severed the economic relations that had been established during the Soviet era, greatly damaging the economy of each newly born country. The transition to private ownership from planned regulation confused many agricultural producers. People who for 70 years were brought up in the spirit of collectivism were initially at a loss. The market economy introduced in the sovereign Kyrgyz Republic has put poor producers—private, farming, and peasant enterprises as well as the *kolkhozes* and *sovhozes*—in a difficult position.

With a view to improving this situation, the Parliament of the Republic passed the laws "On the land code," "On land reform," and "On peasant farming" in 1991/92. In accordance with these normative-legal documents, the privatization of the lands of *kolkhozes* and *sovhozes* should have started with lands that were not cultivated but could be used for crop production, with the land allocated on the basis of competitive proposals from individuals. Similarly, as effective project and program proposals came in, other lands could also be used. Each allocated plot of land would be assigned livestock, poultry, and equipment. First, unprofitable farms would be privatized, followed by farms with profitability below 15 percent. One-quarter of lands—pedigreed stock farms—would not be privatized. They would remain state property for solving any general agricultural problems in the republic.

Such a scenario would have facilitated the expansion of the area of arable land and attracted the more prepared and propertied layers of the population to the land use. The agrarian reform should have included the gradual formation of the farming and peasant enterprises and their associations and joint stock companies. The private owners of land and other immovable property should have attracted other segments of the population on the basis of their positive experiences. The small enterprises as well as the farming and peasant enterprises should have been exempt from some types of taxes for two years.

Unfortunately, violations occurred while implementing these laws because the parties responsible were also the managers of these farms. The farm managers in most cases did not give the land and livestock to those willing to leave the peasant farms. Even if the land and livestock were released, it was done through cronyism and nepotism, and the recipients did not have experience in agriculture and animal breeding. Also the average share of each citizen was not taken into account. Thus, the collective and state assets were squandered. Many farms formally turned into the associations and joint stock companies of the cooperatives, peasant farms, and small enterprises.

An unhealthy atmosphere arose on many farms as a result of this biased approach to the privatization of lands and other immovables. At the same time, production was undermined by propaganda idolizing the private owners and slandering the *kolkhoz* and *sovhoz* systems of production. The government loosened its control over managers in charge of these production systems. Credit was extended irrationally, without any purpose or collateral. When lands were given to industrious peasants, they achieved high crop yields; but when land was released to dilettantes, it remained uncultivated and farmers received low crop yields. Many large farms remained ownerless. Most of the livestock was stolen and slaughtered for various reasons, and the system of veterinary service collapsed. Consequently, the reform did not have the expected result, and agricultural output fell by 15 to 20 percent annually.

During this period, the productivity of fields and farms decreased. In 1995 the average yield of grains was only about 18 centners per hectare, and many fields remained uncultivated. The population of sheep—the main livestock—decreased by half. The stock of poultry and pigs fell by 70 to 80 percent. Overall productivity and fertility decreased, and disease and death among the livestock increased (Tables 12.2 and 12.3).

The republic needed grains and other foods. Unlike the first phase, during the second phase of the reform process, the lands and the property of the *kolkhozes* and *sovhozes* were subject to privatization carried out by the newly formed agricultural committees and the Ministry of Agriculture and Foods. However, the land belonging to the National Fund was still owned by the government. The government provided that specialized farms engaged in producing highly productive seeds and the breeding stock and commodity-producing enterprises would be formed on these lands even though they would remain on lease terms for the short run.

At the same time, the reform provided for the voluntary formation of household and farming enterprises, peasant farms, and associations and cooperatives. Upon the decision of the collective, the collective forms of farm management may remain. The scientific and research establishments, the experimental pedigreed stock farms, and the seed growing farms will remain the property of the government. Joint stock companies and governmental joint stock companies will form

TABLE 12.2 Yield of agricultural crops

Product	Years				
	1981-85	1986-90	1990	1995	1998
	(tons per hectare)				
Grain crops	2.5	3.2	2.7	1.8	2.8
Cotton	1.9	2.5	2.7	2.2	2.5
Potato	14.0	13.7	10.0	9.9	13.1
Tobacco	2.5	2.6	2.2	1.8	2.2
Vegetables	20.2	20.9	18.0	13.2	14.3
Cucurbitaceous crops	12.7	13.5	13.7	12.3	12.2
Sugar beets	20.9	No data	15.6	14.2	18.1

SOURCE: National Statistics Committee.

joint ventures with representatives from other countries for running the poultry farms and other livestock breeding complexes and processing enterprises.

During 1994/95, land and property privatization progressed throughout the country. About 25,000 farming and peasant enterprises, 680 cooperatives, and 237 associations of peasant farms and joint stock companies were created. Technical service centers supporting the farms transformed the machine-tractor parks and workshops, and veterinary service centers emerged from the former veterinary points. Processing enterprises were privatized and became joint stock companies and government joint stock companies. Small enterprises engaged in all spheres of agricultural production such as procurement, storage, processing, and sales of agricultural products including vegetable oil, potatoes, and vegetables. The area primarily under wheat, potatoes, and vegetables expanded sharply to the detriment of fodder crop area.

However, during this period, livestock production continued to decrease sharply (Table 12.3). People preferred to keep horses and milk cows. The breeding composition of the livestock got worse because of the breakdown of the sys-

TABLE 12.3 Livestock production during perestroika

	1985	1990	1995	1998
Production				
Meat (thousands of tons)	169	451.1	321.4	329.8
Milk (thousands of tons)	771	1,185.0	864.2	972.7
Eggs (millions of pieces)	532	713.8	146.7	175.8
Wool (tons)	32,500	39003	147.3	11441
Population (thousands of head)				
All cattle	1,110	1,205.2	869.0	910.6
Cows	427	506.1	476.9	492.2
Sheep and goats	10,200	9,972.9	4,274.8	3,810.6
Female goats	6,840	5,942.2	2,601.6	2,297.6
Pigs	349	393.1	118.1	105.5
Horses	385	312.6	308.1	335.2
Poultry	12,394	13,914.6	2,031.8	2,727.5
Productivity				
Milk yield per cow (liters)	2,789	3,302	1,750	2,053
Wool clipping per sheep (kg)	2.8	3.8	3.0	3.4
Calves born to 100 female cows	98	82	64	80
Lambs, kids born to 100 female goats	102	95	87	94
Foals born to 100 female horses	72	66	63	71

SOURCE: National Statistics Committee.

tem of raising pedigree livestock. Together, these decisions resulted in a crisis in the food industry. The operation of all enterprises functioning earlier decreased by 50 to 80 percent, depending on their financial security and the supply of raw materials. In 1996, the gross production of the agricultural sector fell by 2 percent compared with 1994. In other production areas the change was not significant.

The comforting news is that the overall decrease in production of the agricultural sector in Kyrgyzstan has begun to improve. However, the people of Kyrgyzstan became politically minded—as they did during 1918-36—and instead of fulfilling the plans set by the reforms and improving production, they wasted their strengths and time fighting nonexistent enemies. The unclear position and actions of the government on issues related to privatization and production contributed to this. These issues gave many farm managers and heads of *rayons* (subprovince-level administrative units) the opportunity to make "mistakes" and abuse the privatization process. The government did not properly manage its share of lands, did not adequately finance the property, and did not demand increased production efficiency from those in charge.

Experience has shown that in a transition period, the implementation of agrarian reform should be carried out without loosening the personal responsibility of farm managers. Each type of enterprise, whether state, collective, or private, should work rationally and efficiently for the well being of the whole population and with the support and control of the government. Unfortunately, this has not happened in the Kyrgyz Republic.

In our opinion, full-fledged "democracy" within such a large-scale economic transformation can only exist in rich countries with a well-educated population. In poor countries, the government must control everything in order to increase the food supply. We believe that the new countries that emerged after the collapse of the Soviet Union are among such countries. The republics that had become accustomed to managing their agricultural sectors on an industrial basis are now having difficulty transforming into systems based on horse and manual technologies on small plots of lands. Large losses in the agricultural sector—in immovable capital, in particular—have resulted in a poor food supply. Sheep pens, cattle-breeding complexes, and large processing enterprises were demolished. Consequently, the society became demoralized, the existing seed-growing enterprises and pedigreed stock farms worsened, and the quality and quantity of the general funds of plants and animals decreased.

Owing to the stimulation of the agrarian reforms in the Kyrgyz Republic, certain improvements occurred during the second phase in agricultural production and supplying the population with food products. Most important, in 1996 the gross production of grains, vegetables, and potatoes was stabilized. This was primarily due to the expansion of area under these crops and some improvements in productivity compared with 1994-95. Farmers in Kyrgyz Republic started planting sweet beets and expanding the area under oil producing plants; new interest in increasing the production of raw cotton and tobacco emerged; and the decline in milk production stopped.

The livestock and poultry population declined until 1998. However, because peasants and farmers did not develop the skills needed to raise pedigreed stock, the general fund of animals continues to worsen. This, accordingly, results in poor productivity. In the agroindustrial complex, the processing sectors are still in crisis. Decreases in the milk, vegetable, and meat processing sectors and some light industry sectors continue.

In one new development, people are changing their attitudes toward investors and credit and are now willing to take measures to increase the efficiency of their management. Having felt these positive prospects, the President and the Parliament in 1999 passed four new laws by referendum that aim to boost the efficiency of the agroindustrial complex in the third and final phase of the agrarian reform. The revised law "On the land code" provides for the transfer of lands to peasants free of charge, with the subsequent permit to buy and sell. The law "On the hypothecation" allows a landowner to obtain credit by pledging his plots of land to a special mortgage bank. The new law "On cooperation" permits peasants to associate and form peasant farms, cooperatives, and other types of collective farming at their discretion and to manage their farms on an industrial basis. Finally, the adoption of the laws "On seed farming," "On raising the pedigreed stock," "On veterinary science," "On the development of small and medium-size enterprises," and others will gradually improve the work in these areas. The republic took these steps to decrease land taxes and to promote the high achievement in the agroindustrial complex.

New laws geared toward the rational use of labor and natural resources with a view to ensuring the sustainable development of agriculture, processing, and other sectors of the agroindustrial complex were also passed. Finally, scientific approaches are being introduced to improve the general fund of plants and agricultural animals and to protect the environment of the Kyrgyz Republic. Nevertheless, according to data from the National Statistics Committee, the annual consumption of food products per capita is still low (Table 12.4). This deficit in the consumption of many types of provisions is compensated in part by the consumption of floury products.

At present, more than 65 percent of the population of the republic lives below the poverty line. The analysis of the transition from the socialistic, planned system to one based on market relations has shown that in the agroindustrial complex this task should be carried out very carefully, taking into account the current situation and the mentality of the country's population.

Unfortunately, the factors stabilizing the productivity of fields and farms are used poorly. Each year, between 80,000 and 100,000 hectares of arable lands remain unsown, and the application of fertilizers for agricultural crops has dropped sharply. Most landowners do not combat diseases, pests, and weeds on their lands; and only about 30 percent of the large pasture areas are used.

Because of the excessive fragmenting of lands among the various types of ownership, the crop rotation introduced earlier is now upset, and a new system of crop rotation has not yet been introduced. The irrigation structures and channels

TABLE 12.4 Per capita consumption of food, 1990-97

	Quantity			
Product	Norm	1990	1995	1997
Meat and meat products (kg)	70	54	38	42
Eggs (pieces)	265	154	33	37
Milk and milk products (liters)	290-360	266	172	187
Potatoes (kg)	105	69	82	87
Vegetables and cucurbitaceous crops (kg)	140	78	44	65
Fruits and vegetables (kg)	75	16	7	25.3
Sugar and bakery products (kg)	35	37	15	16
Vegetable oil (kg)	13.5	11	5	4.1
Bread products (kg)	115	132	109	156

SOURCE: National Statistics Committee.

and drainage networks that were constructed earlier are breaking down. As a result of these and other factors, young people are now leaving the villages. All these negative occurrences in agriculture concern the scientific world and the Kyrgyz nation. Nevertheless, at the same time, the following positive events have taken place:

- The nation has begun to use its natural resources rationally and creatively (including land, vegetable, and animal resources).
- The nation has realized the power of market-relations laws and within a short time has moved in this direction.
- The nation has understood that only those who work hard can become rich, and its population has therefore begun to work hard.
- The nation has understood that market laws work best for the literate, and its population has therefore begun to study.

As a result of these events, the following effects have been observed:

- Today in the Kyrgyz Republic, the number of students studying at higher institutions has increased by 50 percent. Large-scale, ongoing retraining is being done to teach fields of specialization.
- According to the National Statistics Committee, one-third of the population of the republic is now migrating both inside and outside the country to find remunerative work. This is something new for the republic.
- More than 50,000 farms, peasant farms, cooperatives, joint stock companies, and a host of other types of small enterprises engaged in the procurement, storage, processing, and sale of products were created in 1999 as a result of the agrarian reforms in the agroindustrial complex.

- A number of small farms and peasant farms with good economic success have emerged. Peasants, having felt the need to combine their efforts and means with a view to achieving profitable commodity production in the sectors of the agroindustrial complex, are forming new types of peasant farms.

Conclusion

To ensure the sustainable development of the agricultural sector, policymakers must understand the interest of the producers of the final output. In our opinion, the transition from a socialistic, centralized planning system to a system of production through market relations must be phased in using civilized methods and without giving in to personal and political temptations. We believe that the transition should be executed as follows:

First, human rights must be observed. If people want to produce and have the necessary material and educational possibilities, then they should be supported and given the opportunity to develop the sector without decreasing the previous level of productivity. If people want to retain the collective methods of farm management based on private or other ownership, then they should be given the corresponding rights and opportunities.

Second, the Kyrgyz Republic has unique natural and climatic conditions, rich fauna and flora, and an industrious and experienced people, all of which should be used to their fullest extent. The problems in the agricultural sector will be solved in the near future by methodically implementing and improving the adopted agrarian policy and taking into account this analysis of the current situation and conditions. In our opinion, this will require the priority development of the sectors of science and new technology and the preparation of highly qualified personnel.

Third, the agrarian policy of the transition period must ensure food independence without a decrease in the volume of agricultural production, regardless of the manner of reconstruction. Reconstruction should be carried out taking into account the mentality of the population and the fact that agriculture is a subsidized sector.

Fourth, any government should act as the donor, sponsor, major creditor, and customer of its country's agricultural products. The food supply must be stimulated through appropriate credit and taxes to ensure accelerated development. The state must not dissolve its enterprises and departments, which are providing village populations with beneficial services, until a sound system of mutual services based on market relations is created within the agroindustrial complex. In particular, the state and the professionally trained collectives must supply scientific support to the agroindustrial complex, must produce highly productive seeds and pedigreed stock, must develop veterinary services, and must control the quality of products.

13 Agrarian Reform in the Kyrgyz Republic

KACHKINBAY KADYRKULOV AND K. KALCHAYEV

The Kyrgyz Republic is a small, mountainous country. Approximately 90 percent of the territory is covered by mountains, and the highest altitude of the mountains is 7,439 meters above sea level. The valleys are mainly located at the altitude of 600-1,400 meters above sea level. The country's climate is temperate.

More than 4.7 million people reside in the republic. Out of these more than 3 million live in rural areas. Out of the total land area, 50 percent can be used for agricultural purposes, including 9.0 million hectares for pasturelands. More than 45 percent of gross income in Kyrgyzstan comes from agriculture, and therefore this sector plays a significant role in the economy of the country. The historic changes that took place at the beginning of 1990s forced the Kyrgyz Republic to fundamentally reform its economy with special attention to agriculture.

Land and Agrarian Reforms

The land and agrarian reform under way in the republic is directed toward forming a market economy and gives commodity producers economic and managerial independence. It also provides for the denationalization and privatization of state and collective property and the creation of a competitive environment and market infrastructure by introducing market relations.

The reform process in the agrarian sector has been carried out through changes in the legal system, through a number of laws, presidential decrees, and decrees of the Government of Kyrgyzstan. The country's lack of experience in transitioning from a socialist economy to a market-based system notwithstanding, the legal base for reforming the sector was created within a short period of time. In addition to the 9 laws, 10 presidential decrees, more than 65 decrees and instructions of the government, and other documents, more than 250 normative and legal acts have been developed and passed since 1991, including the ones regulating agrarian reform.

The presidential decrees, the corresponding government decrees, and the approved regulations executing the agrarian and land reforms have laid the principles of democracy, justice, publicity, voluntary decisionmaking, and freedom in choosing the type of ownership and the production activity. For example, the Decree of the President of September 1, 1998, "On the all-nation discussion of changes and supplements to the Constitution of the Republic of Kyrgyzstan," provided for the introduction of land titles. The citizens who reside in the villages and engage in the agricultural sphere are given the right to receive the shares of lands.

Presently, multiproprietary agriculture has developed in the republic, consisting of various types of farms. More than 55,000 farms with different types of ownership have been formed. More than 52,000 farms have been officially registered, including 22,000 farming enterprises and 30,000 peasant farms. Also, 661 collective farms, including 335 agricultural cooperatives, 281 collective-peasant farms, 45 joint stock companies, and 53 state and other types of farm have been created.

The rural population has been issued 930,000 certificates of various forms of ownership. To date, about 2.5 million people have been allocated land plots, totaling an area of 882,000 hectares. However, many people in rural communities, having received the rights to plots of land, still cannot realize their opportunities for various reasons. The National Land Fund had been established to stabilize the social situation; subsequently it was transformed into the Fund for the Redistribution of Lands, with a total area of 353,300 hectares to distribute.

Before 1997, many state agricultural lands were used inefficiently. This has been partially addressed by the government decree which established some market mechanisms. Before the introduction of the unified mechanism for using land belonging to the Redistribution Fund, the monetary funds amounted to 6-8 million *soms*, which were used without due control and effect. Since the adoption of the government decree, the lands of the Redistribution Fund have been leased out on a competitive basis. As a result, rental agreements for 113.8 thousand hectares for a total of 58 million *soms* were signed in 1997. This is double the amount of previous years. Seventy percent of these funds goes to the *ayil okmotu* budget; 30 percent goes to the budget of the *rayon* (subprovince-level administrative unit) to be used at the discretion of the village and *rayon* for the interests of the local population.

Formation of the Market for Land

The experimental auction of the rights to the land in the Redistribution Fund was conducted with a view to forming a land market. The first auction was conducted in 1998 in the Boo-Terek village of the Bakay-Ata *rayon*. Three lots were for sale with areas of 13, 19, and 40 hectares. Ten applications were accepted for the auction. It was an open auction, conducted publicly and democratically. The starting price of 2,003 *soms* per hectare was bid up to 6,000 *soms* per hectare.

In total, 11 *rayons* held 34 auctions and sold 2,242 hectares of land for 7.6 million *soms*. Unirrigated lands in Tyup *rayon* were sold for higher prices than some unirrigated lands in the United States. Nearly 700 people submitted applications, out of which 653 were admitted to the auctions and 212 won. More than 3,000 people, mainly local residents, observed the auctions. All 34 auctions were conducted as workshops. These auctions stimulated the formation of the land market, and all observers and participants witnessed for the first time the determination of land prices.

Impact of Reforms on the Crop and Livestock Sectors

The administrative center charged with executing the land and agrarian reforms under the Ministry of Agriculture and Water Resources reorganized 408 farms. Reforms executed in the agricultural sector have started to bear fruit, and most rural citizens better understand the agrarian and land reforms, particularly since the all-nation referendum positively addressed the issues related to the introduction of private ownership of land. At the beginning of the reform process in the late 1980s, the all-union fund invested heavily in agriculture. Specifically, thousands of tons of fodder and large quantities of agricultural equipment, fuel, lubricants, fertilizer, and seeds went to support such sectors as animal farming, poultry farming, pig farming and milk production.

During 1991-94, the crop production sector experienced downward trends in the number of cultivated lands (down to 155,400 hectares in 1995), the production of grains (down to 986,200 hectares in 1995), and the yield capacity of grains (down to 17.6 centners per hectare in 1995). This occurred before the major farm reorganization. By 1998, thanks to well-utilized human resources—and despite the shortage of current assets and financial resources—the number of uncultivated lands decreased to 81,000 hectares, the production of grain crops reached 1.7 million tons, and the yield capacity increased to 27.9 centners per hectare. The structure of crop sowing is also changing according to market demands, with an upward trend in highly profitable crops. Notably, the crop yield on private farms is 20 to 30 percent higher than on collective and state farms. On private farms 53.0 percent of the total land under crops is sown with sugar beets, 89.1 percent with potato, and 87.5 percent with vegetables.

As a whole, the livestock sector and its yields decreased after 1991. However, as a result of measures taken in the husbandry and fodder sectors, the population and volume of the gross production of livestock stabilized in 1995. Thus, by 1998, the number of cattle increased 7.4 percent; sheep, 2.5 percent; horses, 12.1 percent; pigs, 19.8 percent; and poultry, 34.2 percent. Milk yield increased 13.7 percent, and the clipping of wool, 10 percent. Meat production increased 0.6 percent; milk, 4.7 percent; and eggs, 11.6 percent. All of these improvements were accompanied by the reduction of unit costs by 1.5 times through the thrifty use of working assets.

Analyses have shown that state farms have earned 2-3 thousand *soms* of income per hectare of arable land, while collective farms have earned 5-7 thousand *soms*, and the private farms have earned 24-27 thousand *soms*. At the same time, private farms produced their products with lower costs. The trend of developing private farms on a national scale has yielded positive economic results and produced an increase in the gross income of the agricultural sector. Thanks to the agricultural reforms and despite the lack of working assets and financial resources, the sharp decrease in production has been halted; gross income has stabilized and has now begun to grow.

Reforms in the Seed Sector

Since 1991, many changes have taken place in the agricultural sector, particularly in ownership structures and production relationships. As a result, a new economic environment with functioning elements of market relations is being formed. The issue of land title has been addressed. However, for subjective reasons, many state seed growing and pedigreed stock farms have not taken part in these historic changes. The last state reassessment of seed growing farms revealed a negative trend in their development. These farms are not following the recommended schemes for scientifically substantiated crop rotation and are irrationally expanding the area sowed with grains.

A number of legislative and normative acts isolated the pedigree stock farms and seed growing farms from the reform process and did not give them the opportunity for timely adaptation. This isolation from reform greatly damaged the systems of pedigreed livestock raising, of state pedigreed stock enterprises, and of seed growing, as well as harming the thousands of village residents living on these farms. Recently, the situation in many of the pedigreed stock farms and the seed growing farms worsened because these farms have been unable to adapt to market conditions and have retained their former production relations.

Achievements

The main achievements of reform include the following:

- Fundamental changes in agriculture without major conflicts
- The formation of a new economic environment with functioning markets, competition, and decentralized and demonopolized power
- An increase in the portion of the rural community owning land and other means of production on their farms
- A halt in the decreases in the volume of gross agricultural production and a shift toward stable growth
- A shift in cropping patterns to comparatively profitable crops, including tobacco and vegetables, and decreasing unit costs of production
- The resolution of issues related to land title
- The initial formation of land markets, financial resources, and credit institutions and introduction of elements of competition and bankruptcy into agricultural production

Future Challenges for Agrarian Reform

Giving rural people rights to land and property is one of the most important aspects of implementing agrarian and land reform. Achieving the desired economic and social effects will require deepening the reform processes and finding solutions to the following issues:

- Protecting the rights of peasants to land and the means of production and improving the legal and legislative base
- Increasing government responsibility for the state of the agricultural sector and the support of agricultural commodity producers
- Introducing effective economic and market mechanisms for the production of food
- Improving cooperation among and integration of private farms and enterprises engaged in the production, processing, and sale of agricultural products
- Strengthening the structures responsible for the privatization and reform processes and for resolving issues related to citizens' rights
- Identifying lands that could be used for agricultural purposes and for pasture
- Improving and expanding irrigation networks
- Developing a tax system and customs policy
- Encouraging the processing of agricultural products on farms
- Building the market's role in the development and production of seeds and livestock breeding stock
- Ensuring the timely provision of information, advice, and applied knowledge to villages, peasants, and farms
- Training personnel for market economy conditions and providing peasants and their associations with necessary infrastructure
- Encouraging competition among service infrastructures
- Stimulating the investment process and ensuring the effective use of funds
- Perfecting the legal base of the state bodies of power and creating stimulating working conditions for state workers
- Achieving the timely and rational solution of conflicts arising in agriculture and providing the guarantor of rights and agreements

Farmers and village residents now face a number of difficulties, such as the high price and shortage of fertilizers, the lack of protection for plants and livestock, decreasing soil productivity, the high price of fuel and lubricants, problems related to product marketing, the lack options for fighting pests, shortages of high-quality seeds, and issues of irrigation and melioration. For these reasons, many private farmers and peasants cannot realize their potential or build reserves. To create favorable conditions for farmers and other commodity producers, they must be provided with convenient marketing outlets and service infrastructure, adequate investment, and increased assistance from state agencies.

The agencies in charge of the reform process in the agricultural sector are the centers for executing the agrarian reform established in 1994. These centers are responsible for the following activities:

- Enforcing rights to land and property shares for the members of collective

farms and workers of state farms
- Resolving conflicts related to property shares and their allocation
- Reorganizing farms and enterprises
- Explaining new types of organizations and production relations
- Protecting the rights of rural citizens who wish to become proprietors
- Forming markets for the means of production
- Forming the necessary infrastructure for servicing farmers and peasants' farms
- Redeeming funds to the public budget
- Managing the state share of the reorganized farms

This list of tasks proves that the reforms of the agricultural sectors have not yet been completed; only the foundation for the relations in agriculture has been laid. Therefore, much purposeful work must still be done to introduce true market relations to agriculture.

Concluding Remarks

The deepening of the reform process and the achievement of the intended objectives require a lot of time and means. With a view to solving the tasks listed above, efforts have been undertaken recently to attract and use the funds of international donors. The following projects have been initiated: the "Development of sheep raising" project, the "Support for ancillary agricultural services" project, the "Irrigation systems restoration" project, the "Water supply" project, the "Regional development" project, and others. Furthermore, projects are being initiated for the development of yak raising, for horse raising, and so on.

The objectives of the "Development of sheep raising" project are to establish private sheep raising farms, create service structures for cattle growing, and develop and support research on pedigreed sheep raising. The "Support for ancillary agricultural services" project has been undertaken to raise the productivity and profitability of agriculture and to ensure the stability of growth of the sector through accelerated execution of agrarian and land reforms. The project plans to provide consulting services and information on the marketing of agricultural products to the rural community. It will be supplemented by scientific research in the area of applied technology, and the supply of high-quality seeds the technology for plant protection. The "Irrigation systems restoration" project is directed toward supporting the restoration and rehabilitation of the irrigation infrastructure in the Kyrgyz Republic, which will increase the productivity of irrigated fields through improved water supply.

All of these measures will, in the very near future, facilitate the formation of an effective agricultural sector in the Kyrgyz Republic.

14 Agricultural Policy Reforms and Food Self-Sufficiency in Turkmenistan

LADO MKRYTICHYAN, ANADJEMAL HALNEPESOVA, AND JANDURDY ATAEV

During the years since its independence, Turkmenistan has developed its own model for the transition to a market economy. The Turkmenistan economy is highly open to the external world. The share of gross domestic product (GDP) exported is relatively high and represents US$367.1 million. The main exported products are natural gas (63.7 percent) and cotton products (28 percent). At the same time, Turkmenistan has a large import demand for light and food industry products and for industrial goods. For instance, in 1994 the share of food products imported was 44.2 percent; this figure decreased in 1995 to 17.7 percent because of government efforts.

The main priorities for policy reform in Turkmenistan are as follows:

- Integrate Turkmenistan into the global economic system and identify the state's position in international division of labor
- Achieve food security and food independence
- Produce sufficient consumer goods for domestic uses and increase their export
- Expand the level of mineral resources and processing of agricultural raw materials

Agrarian Sector Reforms

At the moment, agrarian sector reforms serve to guarantee the well-being of Turkmenistan's rural citizens, who make up 55 percent of the republic's population. More than 40 percent of manpower is involved in the agricultural sector, and the sector's share represents as much as one-third of GDP. Agricultural production and raw materials are very important for the textile industry. However, agricultural production currently faces difficulties due to low yields, inefficient water use, and high water mineralization.

The main goals of the Turkmenistan agrarian reforms are to increase the role of the private sector and to raise water use efficiency and land productivity. Currently, 90 to 95 percent of water in the country is used for irrigation; the remainder is used for industrial and municipal consumption.

The establishment of *dekhkan* communities in 1995 became the starting point for land privatization. A majority of farms in Turkmenistan operate on the basis of internal lease agreements, which is the natural mechanism for gradual land privatization. Land leases are intended to be long-term and can be inherited by family members.

Financial relationships in agriculture have developed substantially. Since the end of 1996, centralized syndication has ended. Credit is now provided from the budget on a promotional basis through the Dekhkanbank and its branches: the Pagtabank and the Gallabank. These agricultural banks receive a small fee for allotting credit and collecting payments but are not involved as true intermediaries. In the long term the government plans to transform the agricultural banks into viable rural financial dealers, including positive actual loan interests.

Food Self-Sufficiency Approach

The current policies of Turkmenistan aim to achieve domestic food self-sufficiency. The main ways to achieve self-sufficiency in grains, meat, and milk production are outlined in the program called "10 years of stability" adopted in 1994. To achieve these goals, a new subsidized fund for agricultural development was established to provide rural producers with funds for basic production.

In February 1999, taxes on agricultural production were temporarily abolished to increase agricultural production and to improve quality for export. Data on the production and consumption of meat, milk, and other primary food products permitted the calculation of ratios of self-sufficiency, as shown in Table 14.1.

Table 14.1 demonstrates the significant shift toward import substitution. Before the collapse of the Soviet Union, Turkmenistan was a food importer.

TABLE 14.1 Self-sufficiency ratios for primary food products in Turkmenistan, 1991-97

	Ratio of production to consumption						
Food product	1991	1992	1993	1994	1995	1996	1997
Bread	0.15	n.a.	n.a.	n.a.	0.85	0.71	0.95
Vegetables	1.54	1.23	1.00	1.30	1.30	1.36	0.95
Melon crops	2.30	1.56	1.38	1.78	1.46	1.80	n.a.
Oil	2.54	1.91	1.82	n.a.	1.71	1.68	1.70
Meat and meat products	0.75	0.72	0.67	0.70	0.69	0.83	0.80
Milk and dairy products	0.61	0.63	0.84	0.88	0.90	1.40	1.54
Eggs	1.70	0.84	0.83	0.95	0.95	1.28	1.37

NOTE: n.a. indicates not available.
Source: Author's calculations.

Between 1991 and 1998, Turkmenistan's position changed from heavy dependence on grain imports to self-sufficiency in major food commodities. Following the breakup of the Soviet Union, debt relations among traditional partners were complicated, and a moderate increase in food self-sufficiency would be a natural economic response. However, the changes over this period also resulted from the policies of the Government of Turkmenistan.

In the mid- to long-term, agricultural industry will be oriented toward production with efficient use of land and water and should bring high added value. High-quality fruits and vegetables, livestock products, and cotton are probable candidates for agricultural industry development. Under the conditions of completely open trade, Turkmenistan has the opportunity to recover and expand its market for fruits, vegetables, and melon crops. The export of leather, skin, and wool is also viable.

Impact of Policy Reforms on Agricultural Production

At present, agricultural production is divided among the following categories:

- *Dekhkan* farms—independent farms of citizens with judicial rights
- Agricultural enterprises run by businesses and institutions, ministries, and agencies
- Private farms, collective orchards, and yards

We will now examine the main production indexes for Turkmenistan agriculture during the period from 1992 to 1998 for agricultural crop cultivation, agricultural production, and livestock. The total area potentially available for agricultural production in Turkmenistan is 40.7 million hectares, with an irrigated area of only 1.7 million hectares. The total area for all crop cultivation in 1992 was 1.2 million hectares. In 1998 that amount increased to 1.3 million hectares. This cultivated area covers 76.9 percent of irrigated lands, meaning it is possible to increase agricultural production in irrigated areas by 23.1 percent.

Table 14.2 shows that the largest shares of cultivated area are occupied by wheat and cotton. In 1998, the area for cotton cultivation was 543,800 hectares, which was 95.9 percent of 1992 levels. Given the decreasing area under cotton, future increases in cotton output will have to come from increases in yield. The area under wheat cultivation in 1998 was 651,100 hectares, an increase of 3.3 percent over 1992. The area for melon cultivation sharply decreased, by 2.8 percent, and for fodder crops cultivation, by 3.3 percent. Private sector production of such crops as potatoes, vegetables, melons, and fodder crops is planned in the future.

Table 14.3 shows that the largest agricultural production by weight is occupied by cotton fiber, followed by grains and legumes. Wheat production increased particularly rapidly during this period. In 1992 wheat production was 372,200 tons; by 1998 it had increased 3.3 times to 1.2 million tons. In 1999 wheat production was predicted to be 1.3 million tons. These increases in production were

TABLE 14.2 Cultivated area on Turkmenistan farms, 1992-98

Crops	1992 All farm types	1992 State farms	1993 All farm types	1993 State farms	1994 All farm types	1994 State farms	1995 All farm types	1995 State farms	1996 All farm types	1996 State farms	1997 All farm types	1997 State farms	1998 All farm types	1998 State farms
							(thousands of hectares)							
Grains and grain-legumes	330.0	327.0	435.3	426.1	598.2	585.2	657.5	639.0	628.4	602.8	572.7	542.7	674.9	633.0
Wheat	195.5	195.3	259.5	256.3	431.6	427.8	552.2	545.0	536.0	529.4	510.6	480.7	651.6	613.0
Cotton	568.7	568.3	579.9	579.6	558.7	563.5	563.5	531.2	530.4	482.1	482.1	482.1	547.4	544.0
Sugar beat	567.0	567.0	579.0	579.0	557.0	557.0	563.0	563.0	530.0	530.0	482.0	482.0	543.0	453.8
Potatoes	3.3	1.4	4.1	1.4	6.4	1.4	6.0	0.6	7.5	12.1	6.7	0.6	5.8	0.02
Vegetables	26.5	21.1	23.3	14.7	25.1	17.3	23.0	15.3	22.1	13.4	18.9	8.0	16.5	3.5
Melon crops	27.1	25.8	18.7	16.2	24.0	18.8	23.5	18.1	25.5	15.2	16.5	9.4	9.5	3.4
Fodder crops	289.5	284.0	265.4	258.4	248.0	237.6	220.1	207.5	193.5	178.5	168.3	149.8	87.1	76.3
Others	2.4	2.2	0.6	0.4	0.8	0.4	1.0	0.8	1.8	0.8	0.8	0.8	0.7	0.2
Total cultivated area	1,247.5	1,230.1	1,324.2	1,296.8	1,461.2	1,419.3	1,494.3	1,444.7	1,405.2	1,443.2	1,443.2	1,266.0	1,193.3	1,341.5

Source: Author's calculations.

TABLE 14.3 Main agricultural products in Turkmenistan, 1992-98

	1992		1993		1994		1995		1996		1997		1998	
Crops	All farm types	State farms	All farm types	State farms	All farm types	State farms	All farm types	State farms	All farm types	State farms	All farm types	State farms	All farm types	State farms
	(thousands of hectares)													
Grains and grain-legumes	722.2	717.7	974.0	954.0	1,106.2	1,079.8	1,109.2	1,074.6	556.4	534.4	760.1	708.7	1,320.5	1,077.8
Wheat	373.2	372.8	508.7	509.0	711.2	705.2	878.7	866.7	453.1	447.7	706.9	655.5	1,245.2	1,055.7
Cotton fiber	1,400.0	1,300.0	1,341.0	1,341.0	1,282.9	1,282.2	1,294.4	1,294.2	435.5	435.5	635.2	-	-	672.3
Vegetables	311.7	237.9	285.6	187.0	386.0	243.3	376.4	249.0	310.0	179.4	241.2	115.7	276.3	199.8
Melon crops	198.8	180.0	176.0	139.4	215.0	158.7	199.2	146.3	189.8	121.1	117.8	60.7	119.2	32.6
Fruits	48.5	19.4	39.5	17.4	42.7	16.3	36.2	19.0	23.9	13.2	25.2	25.2	33.4	11.9
Potatoes	34.5	8.6	31.4	5.3	20.2	4.6	20.5	1.7	21.3	5.1	7.1	0.7	26.8	0.3
Beets (tons)	85.4	28.4	-	-	49.0	49.0	-	-	44.9	44.9	-	-	-	-
Grapes	126.1	90.2	114.0	81.7	132.8	111.1	163.0	130.2	94.0	82.2	126.9	108.4	142.2	108.6
Meat	97.8	46.4	110.4	50.9	107.4	43.1	110.5	39.6	111.2	29.1	110.5	19.1	121.0	22.0
Milk	470.6	188.2	711.0	181.6	715.3	184.6	726.6	156.6	754.8	94.7	755.4	50.2	765.3	32.9
Eggs (million pieces)	292.0	167.6	267.1	132.6	270.2	119.9	270.0	112.4	273.1	48.3	274.1	24.0	269.0	19.0
Wool	16.4	8.7	18.5	7.8	19.2	8.0	19.3	7.5	19.2	6.1	17.9	5.2	18.8	4.5
Astrakhan wool	729.4	519.1	765.1	546.0	742.2	488.5	722.1	421.4	801.9	427.1	667.4	269.3	400.0	275.8

Source: Author's calculations.

achieved primarily through increases in cultivated area. In the future, increased wheat production should result from increases in yield.

In 1992, 1.30 million tons of cotton fiber were harvested; in 1993 this figure increased to 1.34 million tons. However, in subsequent years, the cotton fiber harvest gradually decreased, and in 1996 production reached only 435,500 tons. This drop was caused by sharp decreases in cotton yield. In 1997 cotton production increased again, and in 1998 it reached 707,000 tons. Further increases were expected in 1999 through technological improvements that increase yield.

The private sector component of wheat and cotton production is also growing. In 1992, 0.1 percent of wheat was produced by the private sector; in 1998 this figure was 15.2 percent. Before 1998 cotton was produced mainly by the state sector; in 1998 the private sector share in the overall production of cotton fiber was 4.9 percent. As for other types of agricultural production, in 1998 milk production increased 62.5 percent over 1992 levels; meat production increased 23.7 percent; wool production, 14.6 percent, and grape production, 13.6 percent. Vegetables, melon crops, fruits, potatoes, eggs, and astrakhan wool production decreased during the period examined because of decreases in production by the state sector (Table 14.4).

Income and Expenses of Households during 1995-97

The main goal of economic and social policy in Turkmenistan is to expand human possibilities in society, and to create favorable conditions for long, healthy, and creative lives. During the transition stage of the economic reforms, the system of social security must contain elements specifying the principles, conditions, and standards of employment regulation as well as social guarantees for the population to prevent decreases in living standards. Increases in wages and salaries, pensions, maintenance allowances, and other social payments to families with children are aimed to the social support of the population.

Growth in the incomes of the population and the stabilization of the national currency have led to improvements in buying capacity and increases in the living standard. Consumption of material benefits and services increased 1.3 percent in 1998 compared with 1997. The effects of the social policies and economic changes on the well-being of the people are evident in data from household budgetary research. The average income per capita in the study households was 73,000 manats per month in 1997, an increase of 1.3 percent over 1995.

Private family farms serve as an additional source of food and income. With relatively stable prices for agricultural production, in 1997 the monetary incomes of private farms increased 1.3 percent compared with 1996. This increase can be attributed in part to increases in production by private family and individual farms. For example, in 1997 private farms provided 18 percent of meat and meat products, and rural citizens provided 34 percent; for milk and dairy products it was 52 percent and 85 percent, respectively; for vegetables and melon crops it

TABLE 14.4 Livestock on Turkmenistan farms, 1992-98

	1992		1993		1994		1995		1996		1997		1998	
Crops	All farm types	State farms	All farm types	State farms	All farm types	State farms	All farm types	State farms	All farm types	State farms	All farm types	State farms	All farm types	State farms
	(thousands of head)													
Cattle	1,004.1	386.2	1,104.5	378.0	1,181.5	368.6	1,199.6	340.1	1,155.6	265.5	1,128.5	223.1	1,488.7	159.6
Small ruminants	6,265.0	4,089.2	6,314.1	3,995.6	65.03	3,046.6	6,574.0	3,795.9	6,138.1	3,340.3	5,956.8	3,136.7	6,675.8	2,564.0
Pigs	212.0	205.3	169.4	152.7	128.2	119.8	82.0	74.8	38.3	31.7	31.8	19.2	12.3	86.8
Poultry	6,461.0	4,003.1	6,528.0	3,411.0	5,781.1	2,498.2	4,991.2	1,532.6	4,237.3	624.7	4,234.9	354.3	3,906.1	393.7
Horses	21.2	17.6	22.5	18.3	24.7	18.7	26.3	18.6	27.2	17.3	25.0	14.7	23.1	13.1
Camels	98.4	59.4	101.1	61.8	111.2	61.5	108.0	60.2	105.6	55.1	107.0	52.4	90.0	45.0

Source: Author's calculations.

was 30 percent and 60 percent; and for eggs it was 71 percent and 106 percent.

Actions taken to support the population led to the growth of social payments (pensions, maintenance allowances, and subsidies). Their share in household incomes increased from 8 percent in 1995 to 12 percent in 1997; in rural areas (with the largest number of dependents per household), their share increased from 11 percent to 14 percent. Although expenditures by the population grew in absolute terms, per capita spending remained at roughly the 1995 level. Consumption expenditures—primarily for food—accounted for 89 percent of total expenditures in 1997, a decrease of 2.9 percent from 1995.

Nutrition expenditures occupy the largest portion of consumption expenditures: in 1997 they were 41,000 manats a month per person; in rural areas they were significantly lower—24,000 manats—because of the availability of produce from private farms. While urban citizens acquire 99 percent of their meat and meat products from markets, the comparable statistic for rural people is 68 percent; for milk and dairy products, it is 99 percent and 10 percent, respectively; for vegetables and melon crops, 100 percent and 50 percent; and for fruits and berries, 100 percent and 25 percent.

Increases in welfare and stabilization of prices have influenced overall demand positively. Expenditures for nonfood products in 1997 increased by 1.2 percent compared with 1995. Expenditures for private services (such as personal care, household help, and dining out) increased from 6 percent in 1995 to 8 percent in 1997; expenditures for housing did not change and remain less than 1,000 manats per year per person. Note, however, these are in nominal terms, and inflation of the national currency against hard currency has not been considered. Monetary incomes in all households increased 26.7 percent in 1997 compared with 1995: in urban areas by 28.9 percent, and in rural areas by 23.0 percent. All these data are shown in Table 14.5.

A review of expenditures on wages and salaries reveals similar growth rates in labor payments and monetary income among the urban population, but a higher growth rate in labor payments than in monetary income in rural areas. The growth rate of compensation payments (bonuses) in urban areas is much greater than in rural areas. The largest single share of expenditures is the purchase of foodstuffs; in urban areas it makes up 55.7 percent of overall income, and in rural areas 49.1 percent. This difference is partly due to unaccounted food intake from their own production in rural areas.

In households' food consumption expenditure, the greatest single share is for the purchase of meat and meat products and bread and bakery products. Bread consumption in rural areas is higher by 3.1 percent, but consumption of meat products is lower by 0.4 percent (see Table 14.6).

In the structure of consumption, the share of bread and bakery products in urban areas doubled in 1997. This increase resulted from decreases in potato, vegetable, and fruit consumption, but in rural areas these changes were negligible. In all households, the share of fish and seafood consumption is small; the share of oil consumption has increased significantly. These data are shown in

TABLE 14.5 Annual average income and structure of its use per household, 1995 and 1997

	All households				Rural areas				Urban areas			
	1995		1997		1995		1997		1995		1997	
Stock Type	Thousand manats	Percent of total	Thousand manats	Percent of total	Thousand manats	Percent of total	Thousand manats	Percent of total	Thousand manats	Percent of total	Thousand manats	Percent of total
Labor payment	124.0	77.9	3,285.3	78.8	131.3	79.8	3,788	79.5	116.1	75.8	2,736	77.7
Pensions, maintenance allowances, subsidies	13.2	8.3	332.5	8.0	10.1	6.1	256.2	5.4	16.6	10.8	413.7	11.7
Compensation payments	2.9	1.8	115.5	2.8	3.1	1.9	130.4	2.7	2.7	1.8	99.1	2.8
Property income and real estate sales	0.5	0.3	0.9	0.0	0.3	0.2	0.2	0.0	0.8	0.5	—	—
Other income	18.5	11.6	435.9	10.5	19.69	12.0	587.2	12.3	16.9	11.0	274	7.8
Total monetary income	159.1	100.0	4,170.1	100.0	164.49	100.0	4,762.4	100.0	153.1	100.0	3,522	100.0
Nonfood purchases	44.7	28.8	1,048.70	25.7	37.3	22.8	1,099	23.0	52.4	35.8	1,056.3	31.1
Foodstuff purchases	81	52.2	2,168.2	53.1	96.8	59.2	2,626	55.0	68.2	46.6	1,667.8	49.1
Personal services payment	8.6	5.5	287.6	7.0	9.7	5.9	359.1	7.5	7.4	5.1	209.9	6.2
Other expenses	21	13.5	581.8	14.2	19.7	12.0	689.5	14.4	18.3	12.5	464.5	13.7
Total monetary expenses	155.3	100.0	4,086.3	100.0	163.5	100.0	4,774.2	100.0	146.3	100.0	3,398.5	100.0

Source: Author's calculations.

TABLE 14.6 Per capita household consumption expenses

	All households				Rural areas				Urban areas			
	1995		1997		1995		1997		1995		1997	
Stock Type	Thousand manats	Percent of total	Thousand manats	Percent of total	Thousand manats	Percent of total	Thousand manats	Percent of total	Thousand manats	Percent of total	Thousand manats	Percent of total
Bread and bakery products	3,009	16.8	107,936	22.5	2,856	9.9	154,765	19.4	3,102	27.5	76,416	27.0
Potatoes	1,720	9.6	24,333	5.1	3,506	12.2	43,195	5.4	644	5.7	12,846	4.5
Vegetables and melon crops	2,706	15.2	44,118	9.2	5,258	18.3	83,091	10.4	1,168	10.4	20,391	7.2
Fruits and berries	1,166	6.5	18,716	3.9	2,676	9.3	39,566	5.0	256	2.3	6,017	2.1
Meat and meat products	5,007	28.0	132,875	27.6	7,112	24.7	211,737	26.5	3,739	33.1	84,863	29.9
Fish and fish products	301	1.7	6,685	1.4	748	2.6	16,318	2.0	32	0.3	818	0.3
Milk and dairy products	972	5.4	42,288	8.8	2,177	7.6	98,272	12.3	246	2.2	8,187	2.9
Sugar and confectionery	1,298	7.3	36,564	7.6	1,978	6.9	56,590	7.1	890	7.9	24,373	8.6
Eggs	520	2.9	7,354	1.5	1,247	4.3	17,621	2.2	82	0.7	1,099	0.4
Oil and other fats	483	2.7	36,207	7.5	360	1.3	34,703	4.3	530	4.7	33,209	11.7
Tea, coffee, soft drinks	679	3.8	23,685	4.9	865	3.0	43,054	5.4	593	5.3	15,142	5.3
Total expenses for foodstuffs	17,861	100.0	480,761	100.0	28,783	100.0	798,912	100.0	11,282	100.0	283,361	100.0
Total expenses for foodstuffs	17,861	56.8	480,761	59.7	28,783	62.6	798,912	62.8	11,282	49.8	287,060	55.1
Expenses for meals outside	453	1.4	7,208	0.9	801	1.7	14,098	1.1	242	1.1	3,010	0.6
Expenses for alcohol	1,388	4.4	21,183	2.6	2,047	4.4	33,998	2.7	992	4.4	13,883	2.7
Total consumption expenses	31,441	100.0	805,438	100.0	46,006	100.0	1,272,025	100.0	22,666	100.0	521,385	100.0

Source: Author's calculations.

TABLE 14.7 Per capita household foodstuff consumption

	All households			Urban areas			Rural areas		
	1995	1996	1997	1995	1996	1997	1995	1996	1997
					(kilograms)				
Bread and bakery products	190	170	171	223	234	217	160	135	148
Potatoes	14	13	23	21	18	33	8	10	18
Vegetables and melon crops	86	80	81	97	91	100	80	75	72
Fruits and berries	18	18	20	24	18	24	20	18	18
Meat and meat products	26	23	23	34	29	30	20	19	20
Milk and dairy products	133	117	106	181	121	109	112	116	104
Eggs (pieces)	35	46	43	49	33	44	37	53	43
Fish and fish products	4	1	1	5	2	3	0	0	0
Sugar and confectionery	12	10	11	15	16	16	8	7	8
Oil and other fats	9	8	9	10	8	8	10	9	9

Source: Author's calculations.

TABLE 14.8 Per capita nutrient and caloric value of foodstuffs consumed, 1995-97.

	All households			Urban areas			Rural areas		
Nutrient	1995	1996	1997	1995	1996	1997	1995	1996	1997
					Calories				
Total calories	2,118	1,856	2,016	2,268	2,027	2,155	1,998	1,752	1,954
Protein	54	50	53	51	54	56	51	47	52
Fat	58	43	46	61	46	48	54	42	45
Carbohydrates	340	216	346	366	348	362	330	296	336
Livestock products as a percentage of total caloric value of daily ration	14.4	13.3	12.1	16.3	13.0	12.2	12.7	13.5	12.0

Table 14.7. Studies also illustrate that in 1997 the average daily share of fats per capita decreased in all households compared with 1995 and that carbohydrates occupy the greatest share of caloric value (Table 14.8).

Research shows that Turkmenistan has adequate natural, labor, and financial resources to provide all its population's foodstuffs. At the moment, the republic is completely self-sufficient with grain, vegetables, melon crops, and fruits; the provision of meat and dairy products is being increased, but the provision of seafood is insufficient. The development of the agricultural raw materials processing industry has accelerated.

Conclusions

In conclusion, the most crucial challenges to food independence in Turkmenistan are as follows:

- Improving soil-reclamation, hydro soil-reclamation, and irrigated area drainage systems;
- Controlling rangeland desertification and soil salinization
- Increasing the natural fertility of rangelands
- Introducing intensive crops to agriculture and increasing the productivity of livestock production
- Increasing other productivity rates on the basis of producers' interest

15 Policy Reforms in Kazakhstan and their Implications for Policy Research Needs

ADILYA BAYDILDINA, AYNUR ALISHINBAY,
AND MANSHUK BAYETOVA

Kazakhstan is the second largest republic among the Commonwealth of Independent States (CIS), and its area is greater than that of all the other CIS countries taken together. The area of Kazakhstan is 2.7 million square kilometers (km^2) (272 million hectares); its distance from west to east is 2,925 km, and from north to south is 1,600 km. The most northern points of Kazakhstan (latitude 55°-22° North) are almost at the same latitude as Moscow, Kazan, and Sakhalin Island, and the southernmost (latitude 40° South) at the latitude of Madrid, Istanbul, Baku, and Khonsiu Island.

Kazakhstan is located inside the continent in the center of the Europe-Asia mainland, which affects the climatic conditions of the country. Cold Siberia and very hot Asia, mountainous taiga and desert, vast steppes and mountains covered with eternal glaciers coexist in Kazakhstan. The most vast waterless spaces of the CIS coexist here with the world's largest closed lake, the Caspian Sea, and with the Irtish River, which has an abundance of fresh water. Almost all the territory of Kazakhstan belongs to the largest riverless area on earth.

A low level of precipitation characterizes the valleys of Kazakhstan. Relative humidity is 30 to 50 percent in the north, gradually decreasing toward the south, and not exceeding 5 percent in the desert. In the forest-steppe the average annual rate of precipitation is 300-400 millimeters (mm); in the steppe that falls to 250 mm. In semidesert and desert areas the annual rate of precipitation is 100-200 mm. Particularly low levels of precipitation (less than 100 mm per year) are typical for the Balkhash area, the southeast area of the Kisilkum desert in the Aral Sea region, and the south of Ust-Urt. The influence of large water reservoirs (the Caspian and Aral Seas and Balkhash Lake) results in some increase in the precipitation rates in a very narrow band around their northeastern coasts. In the foothills and mountains, precipitation ranges from 400 to 1,600 mm, and some years on the western slopes of Altai more than 1,600 mm fall. The annual rate of precipitation varies from year to year. The republic encompasses about 85,000 rivers and water reservoirs, with 8,000 of the rivers more than 10 km long.

Large rivers (Irtish, Oral, Ili, and Sirdarya) run mainly along outlying regions of the country. In the summer most of the rivers either dry out completely or are transformed into a chain of overlapping pools. Parts of these rivers disappear, and only during periods of snowmelt do their waters reach some temporary lakes, which then turn into salt beds in summer. Regarding the availability of river runoff, Kazakhstan occupies the next-to-last position among CIS countries. For the country as a whole, the average rate of river water availability is about 20,000 cubic meters of water per square kilometer per year.

Many lakes are dispersed over the vast space of Kazakhstan, but it is very difficult to identify their number, location, and area. The total water space of the numerated lakes of Kazakhstan is about 49,000 km^2. However, the water area of 90 percent of these lakes is less than 1 km^2. Approximately one-sixth of the lakes are salted.

Kazakhstan is poor in forests; the total area of forests in the country is 8-10 million hectares. The forests mainly play a role in soil protection and water conservation, and for fixing saxaul thickets. Among the CIS, Kazakhstan ranks second after Russia in fishery reserves.

Areas suitable for crops without preliminary reclamation comprise 39 million hectares. In addition, up to 22 million hectares have been identified as suitable for irrigation in the deserts and semideserts, land which has been abandoned for lack of irrigation. Another 70 million hectares of conditionally arable saline land exists for which both irrigation and fundamental reclamation are needed. These areas can be reclaimed, and the costs of irrigation would be recovered rather rapidly if a large enough population settled there. The arable-suitable lands can provide high yields if irrigation is developed and fertilization levels are appropriate. Almost everywhere in the Kazakhstan steppes climatic conditions enable the production of wheat, barley, oats, peas, and sunflowers. In the northern regions rye and polygonum can be grown, and in the warmer southern regions millet, mustard, and alseflax can be grown. In the irrigated lands of the southern desert areas it is possible to cultivate thermophilic crops like rice, cotton, tobacco, sugar beet, and southern vegetables as well as vineyards and orchards.

Areas of rangelands are even more spacious, covering more than 179 million hectares. More than 100 million head of fat-tailed, rough-wooled karakul, and partly fine-wooled sheep, goats, cattle, horses, and camels can be grazed there. More succulent and productive summer rangelands, jailau, prevail in the northern areas with higher humidity as well as in the mountains and foothill areas of central Kazakhstan. In the semideserts, deserts, and winter rangelands thinner grasses prevail. Between these areas are rangelands of the transitional seasons of spring and fall; part of these rangelands can be used in multiple seasons or even year-round. Significant differences in the timing of the seasons in the north and the south vary the timing of the vegetation both between zones and within zones. Similarly, the location of various seasonal rangelands close to each other in the mountain areas favors their use as a year-round feeding base. Snow cover in the valleys is not deep, and during warmer winters livestock can be fattened (if the winter feed stock is ensured).

The climate of Kazakhstan is sharply temperate and extremely arid. A long dry summer with warm storms drying out the soil and a short rainless and snowless winter with frequent thaws are typical for most areas of the republic. Crop farming in some areas is impossible without artificial irrigation. Particularly dry areas are located in central Kazakhstan, the Caspian Sea region, Mangyshlak and Ust-Urt, the Aral Sea region, and Betpak-Daliy. Higher moisture levels are observed in the mountainous and rocky areas of the east and south, and in the northern regions of the republic where rainfed crop farming prevails. Being dependent on rain, vast areas with fertile soils in the south cannot be used for cropping without artificial irrigation. High summer temperatures, high evaporation rates, and extremely low air humidity cause the intensive raising of capillary water (which is often salted), and excessive irrigation results in secondary salinization of irrigated lands.

In the arid and dry steppes, yields of rainfed crops vary dramatically from year to year depending on meteorological conditions, and dry storms occasionally ruin plantations and soil across a million-hectare area. Hence, the successful development of agricultural production in Kazakhstan is impossible without further expanding the irrigated zone.

Historically, the practice of agriculture in Kazakhstan has been defined by natural and climatic conditions. From the 9th century B.C. until the 18th century A.D., only one way of farming existed on the European steppe: nomadic livestock production. The rangelands, the cattle stock, and the population together created a sustainable economic and ecological balance.

The economy of nomadic livestock production was based on conserving resources and nature. After the reclamation of virgin lands in the 1950s, crop farming became the major sphere of agriculture in Kazakhstan; livestock production acquired secondary status because of the reduction of the fodder base. The best rangelands and hayfields were ploughed, constantly increasing the areas under food crops. Most of the ploughed areas were located in zones risky for farming. These regions had fertile soil but a short vegetation period and low precipitation rate compared with even the most problematic grain zones in North America. Nevertheless, the yields of wheat—the major crop in 1989-91—were comparable to those in Canada and the United States because of the use of advanced agricultural methods, the development of an anti-erosion system of farming, the high quality of seeds, and the complete system of fertilizers and pesticides.

The Agricultural Reforms

The collapse of the Soviet Union and reform of Kazakhstan's economy and agriculture turned out to be disastrous for the republic. Many negative factors accompanied the reforms, such as the disintegration of interregional economic relations among the republics of the former Soviet Union, errors made in the privatization of property, and the world economic crisis. Within only a few years, food production, livestock populations, and the material and technical base of the agricultural sector declined to the level of the first postwar Five-Year Plan.

Although the process of restructuring and privatizing the agricultural and industrial systems of the republic is generally considered finished, the goal of developing markets in the agrarian sector has not been achieved because of the failure to create a favorable environment. The organizational and legal forms of production developed after the restructuring of state farms represent either a certain type of collective farming or weak individual farms. Several conditions prevent the development of markets and the transformation of the republic into a developed country:

- The restriction on individual private ownership of land, which is the most important production resource in the agrarian sphere
- The lack of a legal base allocating exclusive rights to owners to dispose of their resources and production
- The general disintegration of the economy and increasing dependence on other countries

The results of the reforms can be assessed in different ways:

- Consider whether the initially formulated goal was achieved and what price was paid to achieve it
- Analyze the changes in the level of economic and social development of the agroindustrial complex.

The second type of assessment is the more objective one. If we take the first viewpoint (achieving the initially formulated goals and objectives), admittedly positive results took place. Institutional transformations have been undertaken in the agroindustrial complex, and there is movement toward developing market relations, allowing business units to choose their activities, and overcoming administrative problems in organization and management methods.

Thus, instead of the 2,500 of state and collective farms in 1991, at present there are more than 62,000 individual farming units, 8,754 production cooperatives, 1,169 business partnerships, 578 joint stock companies, and only 89 state enterprises; there are also subsidiary farming units belonging to different departments. The share of nonstate enterprises is 93.9 percent of all agricultural lands, 94.9 percent of arable land, and 91 percent of livestock and poultry (see Figure 15.1).

The share of households and individual farming units in potato production increased from 60 to 84 percent, in vegetable production from 35 to 64 percent, in meat production from 35 to 65 percent, in milk production from 48 to 71 percent, and in egg production from 32 to 40 percent. In all respects, the private sector became dominant in the agroindustrial complex of Kazakhstan.

Impacts on the Agroindustrial Complex

However, these positive changes fail to reveal the huge losses and crises that resulted from the experiments performed in the agroindustrial complex of the republic. In reality, the reforms turned many efficiently functioning large public farming enterprises into numerous small farms, most of which are not viable

FIGURE 15.1 Land according to landowner category, 1990 and 1997

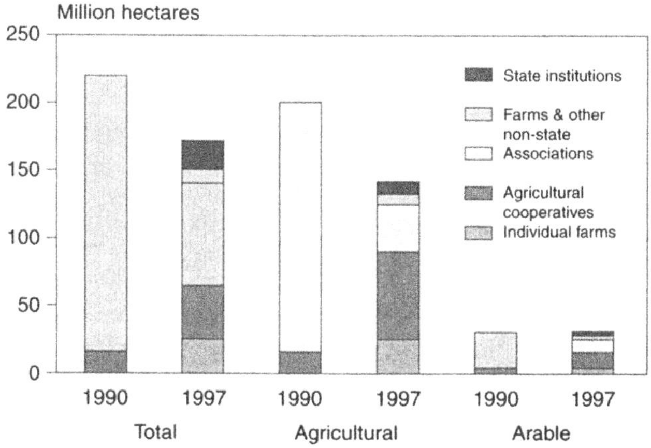

Source: Author's calculations.

because they lack machinery and working capital and cannot adapt to market conditions. As a result, the owners of property and land shares again joined together and established production cooperatives on the basis of joint ownership. Furthermore, social tensions in rural areas have dramatically increased because the protection function under the previous system has weakened.

The areas under agricultural crops are decreasing. Crops covered only 21.8 million hectares in 1998, compared with 35.2 million hectares in 1990. The share of grains (wheat) increased in the cropping pattern while the share of tilled and fodder crops decreased.

In 1997 only 18,900 tons of mineral fertilizers (2.57 percent of the 1990 level) and 163,100 tons of organic fertilizers (0.5 percent of the 1990 level) were applied. No mineral fertilizers were applied in Atirausskaya, West Kazakhstan, or the former Zhezkazgan, Semipalatinsk, and Togaiskaya provinces. The level of fertilizer application is particularly low on individual farms, and the application of chemicals on crops was reduced four- to fivefold compared with 1991. Because farming in Kazakhstan depends on extensive technology, these declines in fertilizer inputs have resulted in reduced yields and affected production sustainability; furthermore, the dependence on climatic conditions is increasing.

Negative effects of the reforms also occurred in Kazakhstan's livestock production sector. From 1990 to 1998, livestock dramatically declined from 18.50 million head to 8.54 million head, or 54 percent (Table 15.1, Figure 15.2). Livestock productivity also decreased. Average annual milk yield per cow fell from 1,985 to 1,558 kilograms (kg), and average annual production of laying hens fell from 229 to 174 eggs. The average weight of one head of cattle declined from 342 to 280 kg, of sheep and goats from 36 to 31 kg, and of pigs from 103 to 54 kg.

The reduction of crop area and population of livestock and poultry, and the

TABLE 15.1 Number of livestock and poultry, all farm categories

Type	Number		1998 as a share of 1990
	1990	1998	
	(thousands)		(percent)
Beef cattle	9,757	4,307	44.1
Cows	3,368	2,110	62.6
Pigs	3,224	879	27.3
Sheep	35,661	10,384	29.1
Horses	1,626	1,083	66.6
Poultry	59,899	15,982	26.7

SOURCE: Author's compilation from unpublished official statistics of the Ministry of Agriculture

reduction of crops yields and livestock productivity have resulted in a dramatic reduction in total crop and livestock production (Table 15.2) and, consequently, in the food consumption of Kazakhstan's population (Figure 15.3).

The reduced agricultural output has increased Kazakhstan's dependence on food imports and weakened the country's food security. Food import costs deplete the gold and hard currency reserves of the country, which have lost their source of replenishment because of the world energy crisis. Food import costs also result in the loss of foreign exchange reserves, threatening the national currency exchange rate and making inflation expectations uncertain. Uncontrolled import of food products hurts local producers and threatens the whole agrarian sector, the share of which in 1997 was half of that of 1991.

Impacts on Prices and Profitability

One of the reasons for the crisis in the agrarian sector can be traced to price liberalization, which resulted in a relative increase in prices for machinery, fertilizers, and energy resources compared to the prices for agricultural outputs. Thus during the reform years (1992-95) prices for industrial production and technical services increased almost 27,000 times, while those for agricultural production increased only 3,948 times. Table 15.3 shows some price changes for specific commodities between 1990 and 1996.

One harvesting machine costs as much as 416 tons of grain or 90 tons of beef, which is 41 to 60 percent more than in 1995, and 55 times more than in 1992. To buy 1.0 ton of fuel, producers of agricultural commodities must sell 2.0 to 2.5 tons of grain. These factors dramatically increase the cost of agricultural products. The growth rates of these costs are 2.4 to 3.3 times higher than the

FIGURE 15.2 Dynamics of livestock and poultry numbers, all farm categories, 1991–97

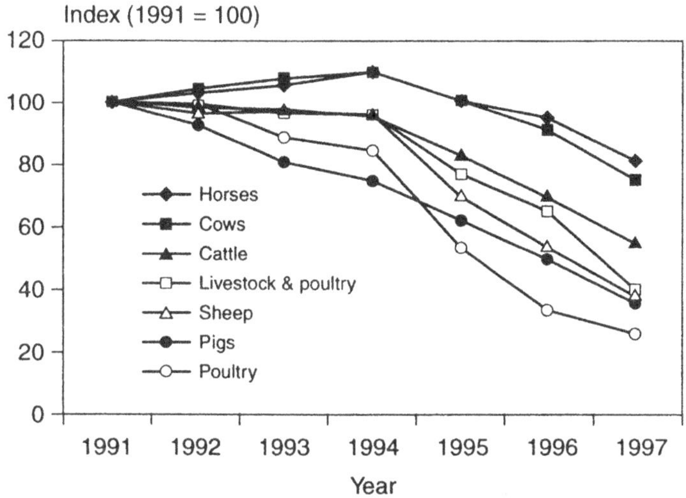

Source: Author's calculations.

TABLE 15.2 Agricultural production in Kazakhstan, all farm categories

Product	1986-90 average	1997	1997 as a share of 1986-90
	(thousands of tons)		(percent)
Grains	24	12	51
Sugar beets	1,433	128	9
Sunflowers	117	55	47
Potatoes	2,114	1,472	70
Vegetables	1,229	880	72
Meat	1,463	718	49
Milk	5,350	3,335	62
Eggs (million pieces)	4,185	1,266	30
Wool	103	34	33

SOURCE: Author's compilation from unpublished official statistics of the Ministry of Agriculture

growth rates for food product prices.

As for the primary cost structure, the share of the vitally important element—the remuneration of labor—has also decreased. Producers of agricultural commodities bear 70 percent of the food production costs but receive only 20 to

FIGURE 15.3 Adequacy of Kazakhstan food consumption, 1990 and 1996

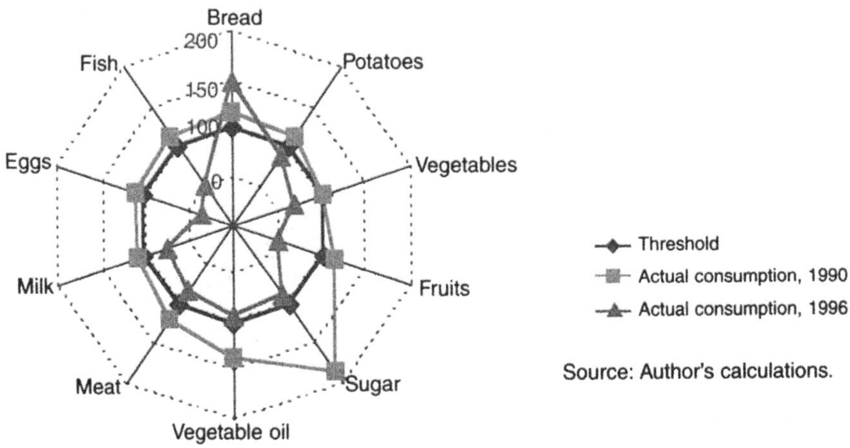

30 percent of the final price for these products.

All these factors resulted from the state's withdrawal from pricing policy regulation. Crop farming and livestock production have become net-loss sectors. Profitability levels in livestock production vary from -1.6 percent to -63.2 percent. As for crop farming, the production of grains, sunflowers, and potatoes remain profitable with a profitability level of +1.7 percent, +49.1 percent, +6.7 percent, respectively (Table 15.4). The state grain bills program that envisaged purchasing grain harvested in 1996 for the state grain stock has negatively affected the profitability of this sector. Because they were financially unsecured, the grain bills were not repaid in time and had to be sold at a discount of 35 to 55 percent. In 1997, prices for type III wheat were US$70-80 per ton (for export, US$110-120), while producers' costs have exceeded US$100.

Agriculture has become an unprofitable sector. The share of farms suffering losses increased from 4 percent to 75 percent (Table 15.5). A dramatic drop in state financial support and loans, the unequal exchange between industrial and agricultural sectors, and the lack of tax privileges resulted in a difficult situation. Almost all producers of agricultural commodities now have a nonliquid balance for their production and finance activities. At the beginning of 1997, their debts exceeded 111 billion tenge.

The current liquidity rate is 0.66, while the standard liquidity rate was 2.0. Producers have negative security in terms of their own funds, but the share of borrowed funds has decreased from 56 percent in 1993 to 44.3 percent in 1996 because of high interest rates. Wear and tear on agriculture's material and technical base is approaching the critical threshold.

Agricultural sector reforms have focused mainly on changing the forms of property, on price liberalization, and on state withdrawal from economic devel-

TABLE 15.3 Increases in production cost of major agricultural products

Product	Cost increase between 1990 and 1996 (multiples): Cost of production	Sale price	Production cost increase as multiple of sale price increase
Grain	27,358	10,270	2.7
Milk	24,249	10,006	2.4
Beef	10,882	4,912	2.6
Pork	25,140	10,697	2.4
Mutton	10,559	3,789	2.8
Wool	7,268	2,199	3.3

SOURCE: Author's compilation from unpublished official statistics of the Ministry of Agriculture

TABLE 15.4 Profits (losses) in the Kazakhstan agricultural sector, 1995 and 1996

	1995 profit (loss)		1996 profit (loss)	
Product	Million *tenge*	Percent	Million *tenge*	Percent
Grains	(921)	(3.8)	658	1.7
Sunflowers	274	107.2	130	49.1
Sugar beets	(1.6)	(18.5)	3.2	(58.3)
Potatoes	147	10.9	90	6.7
Vegetables	(48)	(3.2)	(199)	15.0
All crops	(955)	(2.9)	(923)	(1.9)
Meat	(6,124)	(36.4)	(8,412)	(48.4)
Eggs	(75)	(2.3)	(73)	(1.6)
Milk	(3,630)	(29.9)	(4,365)	(43.8)
Wool	(1,634)	(52.2)	(1,269)	(63.2)
All animal products	(12,601)	(39.7)	(17,915)	(43.7)

SOURCE: Author's compilation from unpublished official statistics of the Ministry of Agriculture

opment. As a result, monopolists dealing with the allied branches of agriculture have continued the trend of placing the basic input costs on the farmer while increasing their own profit share. The system of market regulation is not functioning at the super-monopoly level.

The result has been an almost complete decline in the agronomic, zootech-

TABLE 15.5 Financial status of agricultural enterprises in Kazakhstan, 1990-96

Year	Profit (loss) (million tenge)	Percentage of units suffering losses	Average profit (loss) Total	Crop production	Livestock production
1990	9.9	4	46.0	105.4	18.9
1991	12.2	12	40.2	87.8	24.9
1992	188.1	13	75.8	151.7	(4.9)
1993	(82.2)	51	(2.2)	28.2	(27.7)
1994	(642.9)	62	(6.7)	9.8	24.6
1995	(21755.3)	78	(17.9)	(2.9)	(30.7)
1996	(17249.5)	76	(20.8)	(1.9)	(43.7)

SOURCE: Author's compilation from unpublished official statistics of the Ministry of Agriculture

nical, and veterinary services demanded by the producers of agricultural commodities. The application of mineral fertilizers has decreased by 98 percent, and the stock of machinery is not being updated. In suburban areas, farmers refuse to use trade intermediaries, preferring to deliver and sell their products themselves. To increase their profit, farmers produce processed products by themselves at home, specifically sunflower, sausages, sour cream, and other food products. This has resulted in unused capacity in the processing and resource producing sectors. The rate of capacity utilization for meat processing in 1997 was 17.4 percent, for poultry meat 5.7 percent, for sausages 10.3 percent, for all milk products 13.8 percent, for animal oil 8.1 percent, and for vegetable oil 30.3 percent. (See Table 15.6 for data on production by the food processing industry.)

Many enterprises ceased to exist as whole production units, among them the Semipalatinsk meat combinat, and the Tselinograd amalgamation for anti-erosion machinery production. The Pavlodar tractor plant is seeking customers for its products abroad. Elevator and storage use are at 17 to 18 percent of their installed capacity. Initially, the plan was to transfer storage, processing, and supplying enterprises to collective ownership by the producers of agricultural commodities. However, by 1993-96 the debts of agricultural enterprises reached such a level that when they were divided into individual farms, their property shares turned out to be significantly lower than their debt shares. Thus, the individual farmers had to conclude concession agreements on their land and property with the owners of big financial assets. Rural production is now oriented toward the use of raw materials. Chasing temporary economic success, individual farmers are producing less labor-intensive crops like grain and oilseeds, which hampers fodder pro-

TABLE 15.6 Production of basic food products by the food processing industry

	1990	1995	1996	1997
		(thousands of tons)		
Granulated sugar	319.1	96.6	120.1	147.9
Lump sugar	114.2	16.0	23.0	0.06
Total meat	898.6	273.1	172.9	156.5
Beef and veal	402.4	169.6	111.2	104.9
Mutton	157.0	42.8	28.0	19.8
Pork	142.7	19.4	11.7	11.8
Poultry meat	133.2	23.4	9.6	6.9
Other meat	38.5	8.1	4.1	3.0
Prefabricated meat	154.9	3.1	1.5	0.9
Fish and seafood	85.9	44.6	44.3	40.2
Animal oil	85.1	30.4	15.3	6.9
Whole milk products	1,470.0	278.8	249.5	202.8
Cheese and soft cheese	35.2	11.6	8.6	6.0
Vegetable oil	95.0	43.5	40.5	28.8
Margarine products	71.4	2.8	1.6	5.5
		(millions of tons)		
Total preserved foods	441.8	81.0	55.2	43.2
Preserved meat	38.1	41.7	15.2	4.9
Preserved fish	70.4	9.7	12.0	11.7
Preserved vegetables	74.8	7.7	5.8	6.0
Preserved tomato	102.5	11.1	9.7	4.1
Other preserved foods	156	10.8	12.5	16.5
		(tons)		
Jam	83.4	10.4	12.3	16.5
Juices	0.5	4.8	10.0	8.8
Dairy	0.1	0.4	0.1	0.0
Others	238.0	0.0	0.1	0.0
Dry fruit	258.9	115.0	295.0	124.0
Confectionery	121.9	29.3	n/a	48.7
Saccharine	n/a	12.5	39.6	27.7
Pastry	132.5	78.8	19.0	60.6
Soft drinks (millions of decaliters)	24.1	2.6	68.5	10.5
Mineral water (millions of decaliters)	3.3	1.4	8.4	n.a
Salt (exploration)	n/a	313.0	1.8	3.1

SOURCE: Author's compilation from unpublished official statistics of the Ministry of Agriculture

Note: n.a. indicates not available

duction. The state refusal to finance and provide credit support to producers of agricultural commodities has only increased the price disparity. Prices for industrial products exceeded prices for agricultural ones by 2.4 times in 1993. The sale price of agricultural products in 1997 amounted to 109.6 percent of the previous year, and the industrial products and services acquisition index of agricultural enterprises for the same year amounted to 114.7 percent.

Environment

Conservation of the environment, particularly soil fertility as a basis for agriculture, is an important geopolitical task. From the macroeconomic viewpoint an alarming situation is developing in Kazakhstan: current production is far below that of previous generations, and the established potential for production is being damaged. The excessive introduction of large-scale crop farming in Kazakhstan has undermined the balance between people and nature that existed for thousands of years.

When the fodder base decreased following independence, farmers began to use unsustainable land management techniques to provide fodder for their livestock. One of these techniques was the so-called mowing of pastures, which precluded the restoration of the natural plant cover. The grazing of livestock, sheep in particular, prevented the vegetative reproduction of plants, and seed reproduction was hindered by herbage mowing before seed maturation. This resulted in such rapid degradation of soils that the land was withdrawn from agricultural operation.

By 1997, significant areas had been withdrawn from agricultural turnover; the area of arable land has decreased by 18.3 percent since 1990. Of the 175 million hectares of surveyed lands, 23 million hectares have been degraded and leached by water erosion, including 2.87 million hectares of arable lands; 13.7 million hectares have been severely damaged, including 22,000 hectares of arable lands. Thus, the fodder base for livestock production in the form of rangeland and hay mowing areas has been degraded quickly and severely.

Land Reform

Land reform is the major component of agrarian policy in Kazakhstan. The essence of land reform is the development of a complex set of legal, economic, and organizational measures securing the transition to a new land tenure system in conformity with the character of market reforms. Analyses of the recent changes to the land tenure system reveal that no radical transformations have occurred. The creation of individual farming units has not been attractive for most agricultural workers, and a class of strong private land owners, masters of their own land, has not been formed. The distribution of the plots of land to individual farmers was performed unsystematically, without plans or guidelines, and now the efficient usage of the land plots is compromised.

The country lacks legal arrangements to support the rights of rural working people to their shares of land. By the beginning of 1997, only 60 percent of individual farmers had legal documentation for their land. As a result, the other 40 percent of individual farmers—who do not perceive themselves as the masters of their land—lease their land share rights to managers of farming units without any conditions, and they themselves become hired workers. This phenomenon has occurred in all regions, and in the northern area, 90 percent of individual farmers lease their shares to other persons. Adoption of land laws and establishment of institutions for private property land rights are required to solve the problems in agriculture.

The issue of land tenure is a central problem of reform. The best option, private ownership, wherein the commodity producers have a right and responsibility to use their land and pay certain taxes, is being promoted in Kazakhstan now. Private land owners are entitled to lease their land, use it as collateral in leasing transactions, and transfer it to inheritors. This form of land tenure challenges a centuries-old tradition in Kazakhstan, wherein there was no private ownership of land. However, land reform by itself does not solve the problem of food security, which depends greatly on investments. The slow influx of financing from both internal and external sources is the main reason for the slowing of the production growth rate.

State Support

In their efforts to decrease the state budget deficit, the government dramatically reduced state spending. These cuts amplified the drop in production levels in all areas of the economy. The share of capital investments in agriculture at the national level has fallen by almost 90 percent. According to researchers, the level of state financial support needed for agriculture is high. Investments of US$1.3 billion are needed to restore the production of the major agricultural sectors to the threshold level of food security. This can be justified first by the natural and climatic conditions of the country, as most agricultural lands are located in arid zones. Second, funds are needed to reintegrate the agricultural and industrial sectors for the sake of protecting domestic commodity producers. The state must provide legal and organizational support to achieve parity between these sectors. In this respect, state legislation is often rather late or does not clearly consider the status of the commodity producers. Furthermore, financial support is not always properly addressed. The state should also support research and ensure the research findings and technologies are accessible to individual farmers.

The state must take upon itself a certain part of rural infrastructure development. Establishing a sufficient network of roads is particularly important in Kazakhstan because the population is relatively small and highly dispersed.

In addition to the problems of agriculture production, the food market suffers from a low elasticity of prices; according to our estimations, this index in recent years amounted to 0.04-0.13. Efficient measures are needed in terms of the state regulation of the food market. First, producers should be given funding, preferential crediting and taxation, and insurance. Second, the state needs to protect the interests of domestic commodity producers who are dealing with foreign partners. And third, increases in effective demand must be created through the development of small and medium-size businesses and increases in salaries.

At first glance, the food market appears to be saturated, as there are no queues. However, this is an illusion created by the low buying capacity of the population. Food product price liberalization and growth increased the cost of the food basket, which then became inaccessible for a significant portion of the population. Three-quarters of the population receive rations of bread, potatoes, and milk. The nutrient composition of the ration is primarily carbohydrates; the consumption of fats and proteins is falling; and the diet is not vitamin-balanced. (Tables 15.7 and 15.8.)

The general decline in living standards has most severely affected the least protected rural population. Average wages and their growth rates in rural areas are the lowest in the country. Production declines, reductions in livestock population, and decreases in crop areas have caused increases in unemployment, which in turn resulted in the migration of the rural population to cities where there is unemployment as well. The social infrastructure of rural areas has suffered severely. The native population of the country was originally concentrated in villages, and the economic base of people's livelihood was the agrarian sector. The ethnogenesis and statehood of Kazakhstan directly depends on the development of this sector.

Famine and malnutrition have become an everyday reality for the poor stra-

TABLE 15.7 Average annual consumption of basic food products per capita in Kazakhstan

Product	1990	1995	1996	1997
		(kilograms)		
Meat and meat products	73	52	50	50
Fish and fish products	10.3	4.8	4.6	3.5
Milk and dairy products	311	229	211	196
Eggs (pieces)	225	97	70	69
Vegetable oil	11.2	7.6	7.4	6.5
Vegetables and melons	76	56	52	55
Potatoes	86	70	67	68
Sugar	38.0	18.5	17.6	18.1
Fruit and berries (excluding processing for wine)	23	11	10	9
Cereal products	148	185	185	200

SOURCE: Author's compilation from unpublished official statistics of the Ministry of Agriculture

TABLE 15.8 Sufficiency of basic food products in Kazakhstan

	Consumption		Sufficiency level	
Product	Based on scientifically justified rates	Based on minimum consumption budget	Based on scientifically justified rates	Based on minimum consumption budget
	(thousands of tons)		(percent)	
Meat and meat products [a]	1,291.6	693.1	55	102
Milk and dairy products [b]	6,379.3	3,244.8	47	93
Eggs (million pieces) [c]	4,599.4	2,205.2	24	51
Potatoes [d]	1,575.1	1,291.6	67	82
Vegetables and melons [e]	2,299.7	1,165.6	44	88
Fish [f]	267.8	78.8	21	72
Bread and pastry products [g]	1,496.4	2,441.5	401	246

SOURCE: Author's compilation from unpublished official statistics of the Ministry of Agriculture

[a] Scientifically justified rate of 82 kg/person-year, minimum consumption budget of 44 kg/person-year. [b] Scientifically justified rate of 405 kg/person-year, minimum consumption budget of 206 kg/person-year. [c] Scientifically justified rate of 292 pieces/person-year, minimum consumption budget of 149 pieces/person-year. [d] Scientifically justified rate of 100 kg/person-year, minimum consumption basket of 82 kg/person-year. [e] Scientifically justified rate of 146 kg/person-year, minimum consumption basket of 74 kg/person-year. [f] Scientifically justified rate of 17 kg/person-year, minimum consumption basket of 5 kg/person-year. [g] Scientifically justified rate of 95 kg/person-year, minimum consumption basket of 155 kg/person-year.

ta of the population, whose income is lower than the estimated subsistence minimum. Under these conditions, the state must change its practice of nontargeted fees and subsidies, which are creating an environment for abuse. The state must develop a purposeful program of food security for the poor.

Conclusion

In recent years it has become clear that behind the average indexes of per capita food consumption are hidden large differences in the level and quality of nutrition among the various layers and social groups of the population. Only 8 to 10 percent of families have incomes high enough to afford high-quality nutrition year-round and can consume expensive delicacies, fresh vegetables, and fresh fruit. Families in the lowest group (primarily the rural population, unemployed people, pensioners, and the disabled) are continually malnourished, and their rations consist of low-quality foods such as bread, potatoes, and fats. Therefore, one area of state policy for food supply and trade aims to provide food for the lower income layers of the population. The funds allocated for this purpose target not only the producers of agricultural commodities but also the consumers of these products.

The stocks of agricultural machinery decrease with every passing year. The lack of funds for replacing obsolete machinery restricts the ability of farming units to increase their production. During 1990-97, the number of tractors in Kazakhstan decreased from 220,000 to 108,200; the number of grain combine harvesters decreased from 82,000 to 42,000; and the number of maize combines decreased from 15,600 to 7,300. Enterprises for agricultural machinery construction do not work at full capacity because of the lack of demand for their product. Thus, while 41,050 tractors were produced in the republic in 1990, only 2,058 tractors were produced in 1997.

Using agricultural machinery obtained at very low prices during the privatization process, farming units today can cope with the situation to some extent, but they still have difficulty making ends meet. Additional problems will arise when this machinery stops functioning. Many farmers won't be able to acquire expensive machinery, and then production costs will increase dramatically.

Kazakhstan faces several important institutional challenges:

- Developing the structures for a market economy in rural areas
- Solving the problem of the ownership of agricultural inputs by the producers of agricultural commodities
- Determining the profitability of selling the available land, capital, and labor in the open market

Typically in Kazakhstan, businesses are widely dispersed around urban areas, with large distances between producers and customers. Therefore, the development of both production and nonproduction (particularly market) economic infrastructures is especially important. The state must take upon itself the costs of road construction, transportation, and creating favorable conditions for competition.

Finally, several problems constrain the use of Kazakhstan's natural resources: the difficulty of agricultural work in rural areas, the lack of social conveniences, and inefficiencies in production. As an incentive, the people working in agriculture, particularly in remote areas and zones of ecological disaster, should be given special compensation (subsidies, allowances, etc.).

16 Agricultural Policy Reforms in Tajikistan

BAKHRIDDIN AMIROV

In Tajikistan agriculture is the main sector of the public economy, representing significant production and considerable use of natural resources. The success of the republic's economy in providing its population with food and most of its industries with raw materials depends on this sector. As of 1999, the agricultural sector represented the largest share in production of national income (37.5 percent), and the second largest share in the production of gross domestic product (23.3 percent). More than 20 percent of fixed assets and 42 percent of labor resources have been concentrated in this sector.

Despite these statistics, at present, agriculture is suffering an acute crisis, which is affecting the whole economy of the republic. Domestic production is meeting less than 10 percent of the annual demand for grain, 33.5 percent for meat, 43 percent for milk, 54 percent for eggs, and 87.3 percent for potatoes. Insufficient development of the grain, livestock, and potato production sectors has necessitated importing these products from other states. Furthermore, fruit and vegetable products are exported from Tajikistan even though domestic consumption of grapes and vegetables falls below scientifically justified standards by 10 percent and 8 percent, respectively. Although the Government of Tajikistan has undertaken certain measures to assist the rural population, the crisis is ongoing, and agricultural enterprises continue to work under critical shortages of all types of agricultural inputs.

The situation for agroindustrial processing industries is similar to that of agriculture; and as soon as problems occur in this sector, they are reflected in the level of industrial production. Within the most recent five-year period, the production of cotton fiber decreased by 40.8 percent, vegetable oil by 68.5 percent, flour by 50.0 percent, whole milk products by 74.5 percent, wine and liquor products by 60.5 percent, meat products by 86.6 percent, and fruit and vegetable products by 75.1 percent.

In light of the ongoing state of Tajik agriculture, little hope for rapid or easy agrarian reform is emerging from the crisis. Extraordinary measures aimed at halting the recession of production will be needed to overcome the agrarian crisis. A regulated and socially oriented market economy, based on reasonable and expedient agrarian policy, must be formed. The primary measures that must be implemented are as follows:

- Provide civil peace and strong legislative, executive, and judicial power. There is no hope for any new initiatives or any possibility of resolving a single issue without the elimination of political, interethnic, and social conflicts and of corruption in all social groups.

- Provide sufficient foodstuffs for the country
- Provide social security for the population
- Develop rural crafts and local industry. This will generate additional jobs, and help reduce hidden unemployment and provide needed products, including the supply of competitive products to the external market.
- Pursue flexible pricing, taxation, credit, and finance policy
- Work systematically to improve the structure of the entire agroindustrial complex with diverse forms of property and farming. This is intended to achieve efficiency in production and will target inefficiencies in harvesting and postharvest handling, storage, final processing, and retail sale of products.

Considering the particular importance of the providing grains for the population, a program of urgent practical measures was developed to increase domestic grain production, including rice. This program was approved by a Resolution of the Government, and stipulates an increase in grain production of 853,700 tons by the year 2000. To develop the grain production sector and to raise the production of grains to 1 million tons per year—the level sufficient to meet domestic demand—the area under grain is to be expanded to 400,000 hectares (including secondary sowing); the yield is to be increased to 2.5 tons per hectare.

The following measures are to be undertaken for the support of agricultural development:

- Establish and develop the enterprises of the chemical industry
- Develop the network of enterprises producing spare parts and machinery for agriculture
- Arrange capacity building centers for people employed in agriculture
- Establish and develop structures for the insurance of agricultural commodity producers
- Create commodities exchanges for trading cotton, grain, and other agricultural commodities
- Hold regular fairs to promote agricultural inputs and outputs
- Set up broad networks to provide information, auditing, consulting, and marketing services

The recent performance of tobacco and potatoes illustrates the difficulties facing the agricultural sector in Tajikistan. Last year, only the farms in Penjikentskiy and Aininskiy districts grew tobacco, and their planting area totaled only 4,200 hectares. To increase tobacco production will require the revival of former plantations in the Garm group of districts and in other areas, where production ceased because of shortages of material and technical resources as well as poor markets.

For potatoes, the natural conditions of Tajikistan can potentially solve the issue of self-sufficiency with seed and table potatoes within the coming years.

The Garm and Matchin districts are to become the main potato production zones in the republic. The farms of these districts can produce 20-25 thousand tons of potatoes and deliver at least 10-12 thousand tons to other regions of the republic for seeding.

Agricultural Sector Reforms

Provided that a dependable material and technical base can be secured, Tajikistan has the following opportunities.

Cotton

In 1980, 1 million tons of raw cotton was produced in the republic. All the appropriate conditions can be provided to repeat this volume of raw cotton production. At present, cotton producers work under severe conditions. The material and technical base of cotton production has been undermined almost completely, thus reducing cotton yields from 3.0 to 1.5 tons per hectare. However, the actual area under cotton has not decreased during the past 5-6 years. The drop in raw cotton production results mainly from a dramatic drop in yield. Therefore, the first step toward raising the volume of raw cotton production is improving the material and technical supply for cotton producing farms. The volume of raw cotton production can rise to the level of 800-900 thousand tons.

Tobacco

Government plans stipulate that, in the future, Tajikistan will increase tobacco production to 10-12 thousand tons. This will be accomplished by increasing the area under tobacco to 5,000 hectares-4,850 hectares in Leninabad region alone.

Potatoes

Provided that the political situation in the Garm district stabilizes, it will be possible to increase the area under potato plantations to 9-10 thousand hectares. That will allow an increase in potato production of 20-25 thousand tons. If the private sector contributes 20 to 30 percent of the total volume of potato production, the population of the republic can be provided with sufficient quantities of table potatoes.

To achieve the goals for potato production, a strong seed production base must be established. As a foundation for primary seed production, the government will establish a potato production branch of the Bogparvar scientific and production amalgamation based on the B. Gafurov collective farm in Highland Matcha. The B. Khamdam collective farm will specialize in seed farming of primary potato seed. Further reproduction will be carried out on other farms in Highland Matcha.

Researchers have developed a technique to accelerate potato tuber reproduction and have reduced the duration of the stock seed production cycle from six to four years. This will enable the production of high-quality material and the reproduction of sound seed potatoes. At present, the experiment is being carried out on an area of three hectares. The gross production of stock seed material is expected to exceed 60 tons, and next year will be reproduced in Faizabadskiy, Ganchinskiy, and Shakhristanskiy districts. At the same time, the areas under potatoes will be increased in the valleys during the summer planting. Potato tubers grown from summer plantations can be used as seed material under valley conditions to produce an after late planting.

Vegetables

Tajikistan's soil and climatic conditions allow for the growth of high-quality vegetable crops throughout the year. However, because of the social situation in the republic in recent years, the areas under vegetable crops have dramatically decreased. Instead, grain crops—mainly wheat—have been sown on a significant part of this area. Furthermore, because processing enterprises have ceased, the demand for production of these products has also decreased. High transportation tariffs have also contributed to the decreased production of vegetable crops by reducing the export of these products from the republic.

Despite these difficulties, experience in recent years has demonstrated that individual farming units are producing a higher yield and better quality of vegetable products compared with their collective and state counterparts. Individual farms also manage without bank credit while continuing to pay taxes to the state budget. This demonstrates how expedient it can be to produce vegetable and melon crops in the private sector.

Horticulture

A similar situation exists in the horticulture and viticulture sectors. Fifty-eight thousand hectares of orchards and 37,000 hectares of vineyards currently exist in Tajikistan. However, the volume of fruit and grape production is decreasing every year because of a lack of financial resources for purchasing agricultural inputs and the resultant ill-timed performance of agricultural measures. Nevertheless, the volumes of produced fruit and vegetables continue to meet the demand of the republic's population.

Livestock

As of the beginning of 1999, livestock and poultry production levels were as follows: cattle, 1 million head; sheep, 3 million head; and horses, 67,000 head. The livestock production sector can be restored and developed if

- zooveterinary work and systems of purchasing and importing livestock and poultry breeding stock are properly arranged,
- artificial insemination stations work regularly, and
- the issues of agricultural input and acquisition and development of a strong feeding base are resolved.

According to the plans, within the next five years the republic will acquire 50,000 additional head of cattle, 30,000 head of pigs, and 60,000 head of commercial breeds of sheep. At the present stage, at least 200,000 hectares of irrigated arable land should be covered with fodder crops to establish a strong, secured fodder base. However, winter rangelands have frequently been degraded by inappropriate use (eroded, ruined by motor transport, and planted with grain crops for food production). At present, they are being trampled by livestock belonging to the private sector, used for private hay mowing and livestock grazing, and developed for houses. Furthermore, a considerable part of the rangelands is spoiled with poisonous and inedible weeds. Therefore, resolving the issues of redistribution of seasonal rangelands and rational use of these rangelands has become extremely urgent. Swine production sis to be restored to improve the meat balance of the republic and to restore the production volume of high-quality meat products in the meat processing industry. However, to achieve this, it will first be necessary to provide adequate fodder for the stock and to acquire at least 30,000 head of pigs.

Sericulture

A difficult situation emerged in recent years in Tajikistan's sericulture sector: silkworm breeders were not able to produce an adequate supply of cocoons because of the shortage in inputs needed for this process. Specifically, mulberry leaves on which the silkworms feed were in short supply. Although the yield was not particularly low, other problems with cocoon production still emerged. Customers in Tajikistan refused to buy the cocoons because domestic prices exceeded those offered by foreign sources. To further develop the sericulture sector and increase cocoon production, care should first be taken to ensure an adequate feeding base. Plantations of mulberry trees are to be expanded.

Policies Supporting Land Reform

Since 1990, a legislative base regulating agrarian reform has been enacted. All the laws, decrees, and resolutions regarding agrarian reform are aimed at resolving two issues: (1) the transfer of land and property of agricultural enterprises from the state to collective farming units, and (2) the restructuring of agricultural enterprises.

The Resolution of the Government No. 499 dated October 1, 1993, "On measures for drawing less-favored lands into agricultural production," was an attempt to increase agricultural production. To this end, a special land foundation has been established to help create individual farming units and transfer land to leasing collectives, agricultural cooperatives, and individuals for the production of agricultural crops and planting of perennial fruit trees. Having approved the provision of individual (farming) units, this resolution has given a start to this form of farming.

A number of new farming forms have been established in the republic in recent years. As of January 1999, 10,207 individual farming units had been established, with 716,000 hectares of land allocated to them. Enterprises have been established to lease farming units from the state; and thus far 35 farming units have been leased. Interfarm leasing is widespread: 50,000 collective and individual land leaseholders and 33 agricultural cooperatives control 93,000 hectares of land. In addition, 78,000 individual farmers are producing agricultural crops, mainly grains, on 90,000 hectares of marginal lands and wastelands.

According to the Decree of the President of the Tajikistan Republic No. 342 dated October 9, 1995, 50,000 hectares of land suitable for cultivation have been allocated to the population. Twenty farms have been privatized and transformed into open-type joint stock companies to carry into effect the law "On privatization of state property of the Tajikistan Republic." Previous forms of farming may continue to be used if they continue to be profitable. If a farm is unprofitable or suffering losses it should be restructured into a different, more advanced form.

As of January 1999, the Ministry of Agriculture's system included 358 collective farms, 9 interfarm units, 10,000 individual farming units, 35 leasing farming units, and 33 agricultural cooperatives. A lack of development and poor control of activity in many cases resulted in low efficiency among individual farming units. Taking this factor into account, in October 1995 the government adopted the resolution "On restructuring of collective and state farms and other agricultural enterprises." This resolution established an association of leasing, cooperative, individual farming units, and other organizational forms of farming that conforms to the requirements of a market-oriented economy.

The president's decree "On restructuring of agricultural enterprises and institutions," aimed at accelerating the agrarian reform process, was adopted in June 1996. A working group was established with the participation of officials from the Ministries of Agriculture, Foreign Economic Relations, and Finance, and the State Property Committee, to arrange and coordinate work on restructuring collective and state farms and to render advisory assistance. Similar groups have been established at the regional and district levels. Competent experts visit farming units to give practical assistance. Zonal workshops with the participation of all the targeted persons have been arranged previously and are presently being held.

Appropriate recommendations are being developed and distributed in the regions for rendering methodological assistance to provincial and district groups and specialists on the farms, as well as for providing a universal methodology for collective and state farm restructuring.

The Republican Association of individual farming units, with departments in all regions of the republic, was established in September 1996 to protect the interests of farmers, to render judicial and financial assistance for the supply of agricultural inputs, and to assist in the realization of agricultural outputs. The association, jointly with the international organization, "Kara International," also arranged for the training of farmers in rational farming.

PART IV

Country-Specific and Regional Priorities

17 Country-Specific and Regional Issues and Research Priorities

SURESH BABU AND ALISHER TASHMATOV

Following the presentations and discussions during the workshop, each country team came up with policy research priorities for their country and presented the results of the these group discussions to the plenary session. The key policy research priorities identified by the country-teams are summarized below.

Kyrgyz Republic

- Consider new policy directions for agricultural reforms
- Develop policy on regional water trade and use
- Generate policies for irrigation and drainage
- Set policies for rangeland management
- Develop agricultural technology and input policies

Uzbekistan

- Monitor the process of market reforms
- Consider socioeconomic restructuring
- Develop methods to analyze demand for agricultural commodities
- Create mechanisms to implement rural employment policies
- Develop a system for Central Asian countries to exchange information on their experiences in the policy reform process

Kazakhstan

- Analyze the role of agriculture in the economy
- Study market infrastructure
- Analyze production and technological change in agriculture
- Investigate regional trade integration and cross-border trade

Tajikistan

- Study food self-sufficiency policies and regulation of food markets
- Develop policies for a social security system
- Improve tax policies and implement agricultural tax reforms
- Develop microcredit policies
- Analyze the structure of the agricultural sector

Turkmenistan

- Integrate the Turkmen economy into the world trading system
- Review soil erosion control and the reclamation of degraded lands
- Set policies for increasing irrigation water use efficiency
- Analyze monetary, fiscal, and credit policies for achieving food self-sufficieny
- Establish an information center for World Trade Organization market studies

The countries in Central Asia face similar policy challenges, and policy research results generated in one country can likely be applied—with appropriate modifications—to other countries in the region. Thus, the priority policy research issues relevant for individual countries can be grouped under the following major regional themes.

Research on Market Reforms

Although they are not yet completed, much of what has been done in the reform process since independence relates to land reforms and the creation of production cooperatives. Many policy issues remain largely under-researched, such as those related to pricing outputs, marketing agricultural products, determining exchange rates, developing sectoral approaches to improving agricultural markets, and sequencing the policies needed to generate a dynamic agricultural sector in the Central Asian countries.

Regional Trade

Since the breakup of the Soviet Union, the trading system among the countries in the region has collapsed. Delinquencies in payments for commodities imported from one another and the resultant dearth of foreign exchange has brought food and agricultural trade among them to a standstill. There is an immediate need to examine the possibility of a regional trade arrangement among the region's countries and between the region and the rest of the world. Exchange rate and trade policies that enhance the free flow of food and agricultural commodities should be studied with an eye to achieving food security.

Agribusiness and Postharvest Technology

Although the region is home to high-value crops such as fruits and vegetables, the poor organization of postharvest processes and agribusiness has lowered the quality of processing, resulting in heavy losses in production value. Research is needed to inform policymakers of the benefits of appropriate investment in processing and postharvest technologies. This may result in increased income for farmers and the government through increased exports of processed foods.

Food Security and Agricultural Diversification

Because the low level of foreign exchange reserves has reduced trade, countries of this region have focused on food self-sufficiency or "cereal independence" as the sectoral goal for agriculture, even at the expense of efficiency. Thus, the production of cereal crops to meet local food demands has taken precedence over the production of cash crops such as cotton. Cereal crops are also competing for resources that were traditionally devoted to livestock production. Research is needed on optimum allocations of land and water to crop production. The issue of which direction these countries should follow to attain food security and broad-based economic growth remains an issue for empirical investigation.

Water Use Management

All the countries of the region face serious natural resource problems related to water use efficiency. Generating rules for the allocation of water and the formation of water use associations remains the most fundamental need for managing irrigation water resources. Salinity, water-logging, and drainage problems are also of crucial importance. Research is required on appropriate water pricing and water allocation mechanisms.

Sustainable Use of Rangelands

A large share of the rural inhabitants of Central Asia relies on livestock production for their livelihood. Livestock production largely depends on rangelands, which are being degraded at a high rate. The traditional migratory system of ruminant management is disintegrating, with heavy losses of animals in some countries. As a consequence, the level of poverty in these areas is on the rise. Reversing this trend will require policies that encourage sustainable use of rangelands, provide livelihood for rural households, and reduce their migration to urban areas.

These themes and associate subthemes are summarized in Box 17.1. Other policy research issues that are common to all the countries in the region include

- identifying the appropriate macroeconomic environment;
- developing microcredit policies;
- identifying appropriate social security systems;
- monitoring the process of policy reforms;

BOX 17.1
Summary of regional priorities in Central Asia

1. Market Reforms
   - Develop marketing cooperatives and appropriate institutional structures
   - Monitor food markets and prices
   - Analyze how food and input markets function

2. Regional Trading Arrangements
   - Consider regional trading agreements
   - Identify markets for surplus and processed food commodities
   - Monitor food trade and provide information on international markets
   - Develop an optimal import-export policy

3. Agribusiness and Postharvest Technology
   - Invest in postharvest technologies
   - Develop rural small-scale fruit and vegetable processing
   - Identify markets for processed products

4. Food Security and Agricultural Diversification
   - Identify policy options for increasing producer incentives for agricultural diversification
   - Identify technology research priorities
   - Develop methods of analyzing demand for food products
   - Identify policies for increasing purchasing power
   - Assess the effect of increased food trade on food security
   - Provide rural employment
   - Do research comparing food self-sufficiency and food security

5. Water Use Management
   - Assess degradation of irrigation systems
   - Promote rational usc of water and create water-user associations
   - Set appropriate prices for water for irrigation

6. Sustainable Use of Rangelands
   - Consider land reforms, tenure issues, and property rights
   - Assess rangeland degradation
   - Identify soil erosion problems and develop soil fertility policies
   - Reclaim degraded agricultural lands

- creating mechanisms for information exchange among countries; and
- improving the database for policy research and analysis.

In addition, several other regional issues came up during the presentations. First, it was seen as important to develop a regional vision for these countries' food, agricultural, and natural resource sectors in order to guide future policy research. Second, a mechanism is required for projecting key indicators of food demand and supply and for providing country-level inputs to such an exercise. Finally, there is an immediate need to organize the food and agricultural policy analysts in the region to exchange information on the policy reform process among themselves and with the rest of the world.

Contributors

Aynur Alshinbay is a senior scientist at the Kazakhstan Science Production Institute.

Farruk Aknazarov is rector of the Uzbekistan Irrigation Institute, Tashkent, Uzbekistan, and former director general of the Rural Restructuring Agency.

Bakhriddin Amirov is head of the Accounting and Report Department, Ministry of Agriculture and Water Resources, Dushanbe, Tajikistan.

A. J. Asanaliyev is dean of the Agrarian Faculty, Kyrgyzstan Agrarian Academy, Bishkek, Kyrgyzstan.

Jandurdy Ataev is head of management and marketing in the Agroindustrial Complex Department, Turkmenistan Agricultural University, Ashgabat.

Suresh Babu is a research fellow and head of the Training and Capacity Strengthening Program at the International Food Policy Research Institute (IFPRI).

Adilya Baydildina is a senior teacher at the Kazakhstan State Agrarian University.

Manshuk Bayetova is head of the Department of Economics at the Kazakhstan Agrarian Policy Science Institute.

Hasan Buriev is rector at the Tashkent State Agrarian University, Tashkent, Uzbekistan.

Philippe Chabot is a consultant with the Markets and Structural Studies division at IFPRI.

Marc Cohen is special assistant to the director general, IFPRI.

Shukhrat Egamberdiev is chief specialist, Rural Restructuring Agency, Tashkent, Uzbekistan.

Abdulla Fayzullaev is director general of the Rural Restructuring Agency, Tashkent, Uzbekistan.

Eleonora Gaziyants heads the Department of Economics, Management, and Reforms in the Agro-Industrial System, Uzbek Scientific Production Center for Agriculture (USPCA), Tashkent, Uzbekistan.

Francesco Goletti is a senior research fellow in the Markets and Structural Studies division at IFPRI.

Anadjemal Halnepesova is a senior teacher of the Management and Marketing Department, Turkmenistan Agricultural University, Ashgabat.

Dilshod Hidirov is reform specialist at the Rural Restructuring Agency, Tashkent, Uzbekistan.

Nazarbek Ibragimov is a professor and head of the Department of Crop Production of the Kyrgyzstan Agrarian Academy, Bishkek, Kyrgyzstan.

Abduvokhid Juraev is a professor of economics and deputy minister of the Ministry of Agriculture and Water Resources, Tashkent, Uzbekistan.

Kachkinbay Kadyrkulov is head of the Agrarian Reforms Department, Ministry of Agriculture, Bishkek, Kyrgyzstan.

K. Kalchayev is a leading specialist at the Ministry of Agriculture, Bishkek, Kyrgyzstan.

R. Khusanov is director of the Uzbek Research Institute of Market Reforms of USPCA, Tashkent, Uzbekistan.

Jim Longmire is on the business faculty in the Department of Economics and Resources Management, University of Southern Queensland, Toowoomba, Queensland, Australia.

Sultonali Mekhmonov is a program assistant at the Tashkent State Agrarian University, Tashkent, Uzbekistan.

Erika Meng is a research economist at the International Center for Maize and Wheat Improvement (CIMMYT), Mexico City, Mexico.

Lado Mkrytichyan is deputy minister of Agriculture on Land Reform Issues Ministry of Agriculture, Ashgabat, Turkmenistan.

Altynbeck Moldashev heads the Department of Vertical Integration in the Agricultural Complex, Kazakhstan Scientific Research Institute of Agricultural Economics, Almaty.

Peter Oram is a research fellow emeritus in the Environment and Production Technology division at IFPRI.

Roman Ospanov is an economist at the Agency for Farm Restructuring in the Ministry of Agriculture and Water Resources in Tashkent, Uzbekistan.

Rajul Pandya-Lorch is head of the 2020 Vision for Food, Agriculture, and the Environment initiative at IFPRI.

Per Pinstrup-Andersen is the director general of IFPRI.

Stacy Roberts is a special assistant to the director general, IFPRI.

Otabek Shomuratov is an economist at the Agency for Farm Restructuring in the Ministry of Agriculture and Water Resources in Tashkent, Uzbekistan.

Mekhlis Suleimenov is deputy head of the Programme Facilitation Unit (PFU) at the International Center for Agricultural Research in the Dry Areas (ICARDA), Aleppo, Syria.

Alisher Tashmatov is an associate professor at the Tashkent State Agrarian University and Deputy Director General of the Agency for Farm Restructuring in the Ministry of Agriculture and Water Resources in Tashkent, Uzebekistan.

Rafshan Turgunbaev is an economist at the Agency for Farm Restructuring in the Ministry of Agriculture and Water Resources in Tashkent, Uzbekistan.

Workshop Participants

Kazakhstan

Dr. Aynur Alshinbay
Doctor of Economics, Senior Scientist
Kazakhstan Science Production Institute

Dr. Adilya Baydildina
Doctor of Economics, Senior Teacher of the Accounting, Auditing, and Analyzing Department
Kazakhstan State Agrarian University
Tel: (+993 272) 479 252
329 285

Dr. Manshuk Bayetova
Doctor of Economics, Head of Department
Kazakhstan Agrarian Policy Science Research Institute

Kyrgyzstan

Dr. Nazarbek Ibragimov
Head of the Plant Department
Kyrgyz Agrarian Academy
Tel: (+996 312) 449 273
Fax: 550 548
545 545

Dr. Kachkinbay Kadyrkulov
Doctor of Agricultural Sciences, Head of the Agrarian Reforms Department
Ministry of Agriculture
96 A, Kievskaya St.
Bishkek, Kyrgyz Republic
Tel: (996 3312) 263733
Fax: 621387

Tajikistan

Dr. Bakhriddin Amirov
Head of Accounting and Report Department
Ministry of Agriculture and Water Resources
44 Rudaky St., 734025
Dushanbe, Tadjikistan

Dr. Rakhmatullo Ergashev
Dean of the Economics Department
Tajik Agrarian University
146 Rudaky Prospect, 734071
Dushanbe, Tajikistan
Tel: (+992 372) 242 378
245 809
Fax: 216 226

Turkmenistan

Dr. Jandurdy Ataev
Doctor of Economics, Head of Management and Marketing in the Agroindustrial Complex Department
Turkmenistan Agricultural University

Dr. Anadjemal Halnepesova
Senior Teacher of the Management and Marketing Department
Turkmenistan Agricultural University
62 Gor-Ogly St., 744012
Ashgabat, Turkmenistan
Tel: (+993 12) 410 173
510 502

Mr. Lado Mkrytichyan
Deputy Minister of Agriculture on Land Reform Issues
Ministry of Agriculture
63 Azadi St., 744000
Ashgabat, Turkmenistan

Uzbekistan

Dr. Farruk Aknazarov
Professor of Economics, Director General
Rural Restructuring Agency
5 Mustakillik Square
Taskent, Uzbekistan
Tel: (+998 71) 139 1592
Fax: 139 1588

Dr. H. Buriev
Doctor of Agricultural Sciences, Rector
Tashkent State Agrarian University
4 Universiteskaya St.
Tashkent-140, Uzbekistan
Tel: (+998 71) 637 600
639 600
Fax: 637 600

Dr. Tulkin Farmonov
Doctor of Economics, Deputy Director
Uzbek Scientific Production Center for Agriculture
8 Navoy St., USPCA
Tashkent-140, Uzbekistan
Tel: (+998 71) 417 735

Dr. Eleonora Gaziyants
Head of Economics, Management, and Reforms in Agro-Industrial Systems Department
Uzbek Agricultural Science Production Center
8 Navoy St., UASPC
Tashkent-140, Uzbekistan
Tel: (+998 71) 414 954

Dr. Saidasror Gulyamov
Professor of Economics, Head of Agribusiness and Management Department
Tashkent State Agrarian University
4 Universiteskaya St., TashDAU
Tashkent-140, Uzbekistan
Tel/Fax: (+998 71) 636 455

Dr. Abduvokhid Juraev
Professor of Economics, Deputy Minister
Ministry of Agriculture and Water Resources
4 Navoy St.
Tashkent-140, Uzbekistan
Tel: (+998 71) 410 020
410 042
412 620

Dr. Rasulmat Khusanov
Professor of Economics, Director General
Institute of Market Reforms
28 Drujba Narodov St.
Tashkent, Uzbekistan
Tel: (+998 71) 768 600

Mr. Roman Ospanov
Chief Specialist of Rural Restructuring Agency
Ministry of Agriculture and Water Resources
5 Mustakillik Square
Tashkent, Uzbekistan

Dr. Alisher Tashmatov
Doctor of Economics, Deputy Director General
Rural Restructuring Agency
5 Mustakillik Square
Tashkent, Uzbekistan
Tel: (+998 71) 139 1592
Fax: 139 1588

IFPRI

Dr. Suresh Babu
Research Fellow
IFPRI
2033 K St., N.W.
Washington, D.C. USA
s.babu@cgiar.org
Tel: (+1 202) 862 5600
Fax: (+1 202) 467 4439

Dr. Francesco Goletti
Senior Research Fellow
IFPRI
2033 K St., N.W.
Washington, D.C. USA
f.goletti@cgiar.org
Tel: (+1 202) 862 5600
Fax: (+1 202) 467 4439

Dr. Peter Oram
Research Fellow Emeritus
IFPRI
2033 K St., N.W.
Washington, D.C. USA
p.oram@cgiar.org
Tel: (+1 202) 862 5600
Fax: (+1 202) 467 4439

Ms. Rajul Pandya-Lorch
Head, 2020 Vision Initiative
IFPRI
2033 K St., N.W.
Washington, D.C. USA
r.pandya-lorch@cgiar.org
Tel: (+1 202) 862 5600
Fax: (+1 202) 467 4439

Dr. Per Pinstrup-Andersen
Director General
IFPRI
2033 K St., N.W.
Washington, D.C. USA
p.pinstrup-andersen@cgiar.org
Tel: (+1 202) 862 5600
Fax: (+1 202) 467 4439

CIMMYT

Dr. Erika Meng
Research Economist
CIMMYT
Lisboa 27, apdo. Postal 6-641, Col. Juárez
Mexico City, México 06600
e.meng@cgiar.org
Tel: (+52 5) 726-9091 ext 2127
Fax: 726-7559

ICARDA

Dr. S. Beniwal
Head
PFU/ICARDA
6, Murtazaeva St.
Tashkent 700000, Uzbekistan
tashkent-cac@icarda.org.uz
Tel: (+998 71) 137 52 59
137 52 70
Fax: 120 71 25

Dr. Z. Khalikulov
Consulting Scientist
PFU/ICARDA
6, Murtazaeva St.
Tashkent 700000, Uzbekistan
tashkent-cac@icarda.org.uz
Tel: (+998 71) 137 52 59
137 52 70
120 71 25

Dr. Mekhlis Suleimenov
Deputy Head
PFU/ICARDA
6, Murtazaeva St.
Tashkent 700000, Uzbekistan
tashkent-cac@icarda.org.uz
Tel: (+998 71) 137 52 59
137 52 70
120 71 25

Donor Organizations

Dr. Alexander Kalashnikov
Project Management Specialist
USAID
41, Buyuk Turon St., Rm. 146
Tashkent 700052, Uzbekistan
akalash@usaid.freenet.uz
Tel: (+998 71) 133 18 52
Fax: 120 63 09

Dr. Ken McNamara
Environmental Specialist
USAID
k.mcnamara@usaid.gov

Dr. Peter Reddish
Resident Representative
TACIS
4 Tarasa Shevchenko St.
Tashkent 700029, Uzbekistan
peter@taciscu.bcc.com.uz
Tel: (+998 71) 139 40 18
Fax: 136 06 52

Dr. David Pearce
Resident Representative
World Bank
47, Suleymanova St.
Tashkent, Uzbekistan
Tel: (+998 71) 133 50 02

Dr. V. N. Gnanathurai
Resident Representative
Asian Development Bank
32, Kuloltuprok St.,
Tashkent 700100, Uzbekistan
vgnanathurai@urmadb.uz
Tel: (+998 71) 254 72 87
255 48 25
Fax: (+998 71) 100 13 90

Workshop Agenda

Central Asia Workshop on Food, Agriculture, and Natural Resource Policy Research in Central Asia: Setting the Priorities

July 19-21, 1999
Tashkent, Uzbekistan

Cosponsored by Tashkent State Agrarian University; Ministry of Agriculture and Water Resources, Uzbekistan; and International Food Policy Research Institute, Washington, D.C.

July 19, 1999

Session I (Chair: A. Juraev, deputy minister of Agriculture and Water Resources, Uzbekistan)

9:00-9:15 Welcoming remarks

- A. Juraev

9:15-9:35 Opening remarks

- P. Pinstrup-Andersen, director general, IFPRI

9:35-10:00 Opening speech

- A. Juraev

10:00-10:30 Tea break/photo session

Session II (Chair: A. Juraev)

10:30-11:15 Prospects for Global Food Security: A Central Asian Context

- R. Pandya-Lorch, IFPRI

11:15-12:00 Attaining Food Security in Central Asia: Emerging Issues and Challenges for Policy Research

- A. Tashmatov, TSAU/RRA, and S. Babu, IFPRI

12:00-12:45 Food Policy Research for Improving the Reform of Agricultural Input and Output Markets in Central Asia
- F. Goletti, IFPRI

12:45-1:00 General discussions

1:00-2:00 Lunch

Session III (Chair: Lado Mkrytichyan, Deputy Minister of Agriculture, Turkmenistan)

2:00-2:45 IFPRI's Research Priorities and Linkages to Central Asia
- P. Pinstrup-Andersen, IFPRI

2:45-3:30 Group discussions on regional policy research issues and priorities

3:30-4:00 Tea break

4:00-5:30 Group discussions on regional issues and priorities

July 20, 1999

Session IV (Chair: Kachkinbay Kadyrkulov, Ministry of Agriculture, Kyrgyzstan)

9:00-10:00 Presentation of group discussion results
- Group A
- Group B
- Group C
- Open discussion on regional policy research issues and priorities

10:00-10:45 Policy Reforms and Policy Research Needs in Uzbekistan: Country Paper
- R. Khusanov, F. Aknazarov, A. Tashmatov, R. Ospanov, and E. Gaziyants

10:45-11:15 Tea break

11:15-12:00 IFPRI/ICARDA presentation on Rangelands and Livestock Systems in Central Asia: Policy Research Needs
- P. Oram, IFPRI, and M. Suleimenov, ICARDA

12:00-12:45	Agricultural Policy Reforms and Policy Research Needs in Tajikistan: Country Paper- ■ B. Amirov
12:45-1:00	General discussion on the three papers presented in Session IV
1:00-2:00	Lunch
	Session V (Chair: Rakhmatullo Ergashev, Tadjik Agrarian University)
2:00-2:45	Agricultural Sector Reforms in Kyrgyz Republic and Policy Research Needs: Country Paper ■ K. Kadyrkulov and N. Ibragimov
2:45-3:30	Challenges in Reforming Agricultural and Livestock Sector in Turkmenistan: Country Paper ■ L. Mkrytichyan, J. Ataev and A. Halnepesova
3:30-4:00	Tea break
4:00-4:30	Policy Reforms in Kazakhstan and Their Implications for Future Policy Research: Country Paper ■ A. Baydildina and A. Alshinbay
4:30-5:00	CIMMYT presentation on Wheat Production and Marketing Issues in Kazakhstan ■ E. Meng
5:00-5:30	General discussions on the country papers presented in Session V
7:00-10:00	Reception and dinner

July 21, 1999

Session VI
(Chair: M. Adilya Baydildina)

9:00-9:15	Overview of Country-Level Issues ■ A. Tashmatov
9:15-10:30	Group discussions on priorities for policy research (country teams) ■ Group A ■ Group B ■ Group C ■ Group D ■ Group E
10:30-11:00	Tea break
11:00-12:00	Group discussions on priorities for policy research (continued)
12:00-1:00	Preparation for presentation of summary priorities for country-level policy research
1:00-2:00	Lunch
2:00-3:00	Presentation of summary priorities for country-level policy research

Session VII
(Chair: Francesco Goletti, IFPRI)

3:00-3:30	Central Asia Network for Food and Agricultural Policy Analysis ■ L. Mkrytichyan
3:30-4:00	Summary of the workshop recommendations ■ A. Tashmatov
4:00-4:30	Tea break

Session VIII
(Chair: A. Juraev)

4:30-5:00 Closing remarks

- A. Juraev, deputy minister, Ministry of Agriculture, Uzbekistan
- P. Pinstrup-Andersen, director general, IFPRI
- Vote of thanks, deputy rector, TSAU

Made in the USA
Monee, IL
13 September 2026

62981020R00134